Food Microbiology and Food Safety

Series Editor:

Michael P. Doyle

More information about this series at http://www.springer.com/series/7131

Food Microbiology and Food Safety Series

The Food Microbiology and Food Safety series is published in conjunction with the International Association for Food Protection, a non-profit association for food safety professionals. Dedicated to the life-long educational needs of its Members, IAFP provides an information network through its two scientific journals (Food Protection Trends and Journal of Food Protection), its educational Annual Meeting, international meetings and symposia, and interaction between food safety professionals.

Series Editor

Michael P. Doyle, *Regents Professor and Director of the Center for Food Safety, University of Georgia, Griffith, GA, USA*

Editorial Board

Francis F. Busta, *Director, National Center for Food Protection and Defense, University of Minnesota, Minneapolis, MN, USA*
Patricia Desmarchelier, *Food Safety Consultant, Brisbane, Australia*
Jeffrey Farber, *Food Science, University of Guelph, ON, Canada*
Vijay Juneja, *Supervisory Lead Scientist, USDA-ARS, Philadelphia, PA, USA*
Manpreet Singh, *Department of Food Sciences, Purdue University, West Lafayette, IN, USA*
Ruth Petran, *Vice President of Food Safety and Pubic Health, Ecolab, Eagan, MN, USA*
Elliot Ryser, *Department of Food Science and Human Nutrition, Michigan State University, East Lansing, MI, USA*

Tong-Jen Fu • Lauren S. Jackson
Kathiravan Krishnamurthy • Wendy Bedale
Editors

Food Allergens

Best Practices for Assessing, Managing and Communicating the Risks

 Springer

Editors
Tong-Jen Fu
Division of Food Processing Science
and Technology
U.S. Food and Drug Administration
Bedford Park, IL, USA

Lauren S. Jackson
Division of Food Processing Science
and Technology
U.S. Food and Drug Administration
Bedford Park, IL, USA

Kathiravan Krishnamurthy
Department of Food Science and Nutrition
Institute for Food Safety and Health
Illinois Institute of Technology
Bedford Park, IL, USA

Wendy Bedale
Food Research Institute
University of Wisconsin-Madison
Madison, WI, USA

Food Microbiology and Food Safety
ISBN 978-3-319-88278-9   ISBN 978-3-319-66586-3   (eBook)
DOI 10.1007/978-3-319-66586-3

Printed on acid-free paper

This Springer imprint is published by Springer Nature
The registered company is Springer International Publishing AG
The registered company address is: Gewerbestrasse 11, 6330 Cham, Switzerland

Preface

Food allergies are a significant public health concern, affecting up to 15 million Americans, and there is evidence that the prevalence of food allergies is increasing. Protecting consumers with food allergies requires an integrative approach that involves all sectors of the food industry and engages all who take part in the manufacture, preparation, and service of foods. Success can only be achieved with a broad awareness of the severity of food allergic reactions, a clear understanding of risk factors, and implementation of allergen control best practices. Significant advances have been made over the past two decades in our understanding of food allergens and in the development of control measures to minimize the public health risks associated with them (e.g., issuance of allergen labeling regulations, development of guidelines for managing allergens in food production and food service operations, and establishment of food allergy policies in communities and schools). However, broad and successful implementation of allergen controls is still needed. Many gaps exist, including the lack of hazard analysis and risk management approaches tailored to the needs of specific types of operation, absence of detailed documentation on best practices, insufficient dissemination of available information and resources to stakeholders, and shortage of tools and programs to train staff.

It is with these challenges in mind that a symposium entitled "Food Allergens: Best Practices for Assessing, Managing and Communicating the Risks" was held on October 14–15, 2015, in Burr Ridge, Illinois, to provide a forum where researchers, clinicians, and subject matter experts from the government, the packaged food and food service industries, academic institutions, and consumer groups came together and shared information, discussed current efforts, and recommended ways to address public health issues associated with food allergens.

This book, composed mostly of papers presented at the symposium, provides the most up-to-date information on allergen risk factors and innovative control measures applicable to different segments of the food chain, including manufacturers of packaged food, restaurants and other food service establishments, and at home. Key legislative initiatives that are in various stages of development and implementation at the federal, state, and community levels are also highlighted. The resources presented and experience shared will assist stakeholders in establishing best practices that meet the needs of their specific operations for the assessment, management, and communication of food allergen risks.

Effective dissemination of allergen management information to all stakeholders across the entire food chain is needed. This will require strengthening the nation's education and outreach infrastructure and building expertise on identifying food allergen risks and developing allergen control measures. Enforcing compliance with existing regulations plays an important role in ensuring that allergen controls are effectively implemented. Proper training of state and local inspectors is critical. The information presented in this book will facilitate the development of educational materials and allergen management training programs for food production and service staff, extension specialists, and government inspectors. Consumers and other food safety professionals will also benefit from the information presented which will help them recognize and understand allergen control measures that are put in place across the food chain.

We are indebted to the authors for their efforts and cooperation in the development of this book. We acknowledge USDA NIFA for providing the conference grant (Grant # 2015-68003-23310) in support of the symposium. We also acknowledge the sponsorship of EcoLab, Emport LLC, Grantek, IEH Laboratories, Kerry Group, Kikkoman, Marshfield food safety LLC, Neogen Corporation, Northland Laboratories, PepsiCo, and QualySense. We thank Dr. Charles Czuprynski at Food Research Institute, University of Wisconsin, Dr. Stephen Taylor and Dr. Joseph Baumert at the Food Allergy Research and Resource Program, University of Nebraska, Dr. Steve Gendel (IEH Laboratories and Consulting Group), David Crownover (National Restaurant Association), Susan Estes (PepsiCo, retired), Dr. Jon DeVries (General Mills, retired), and Jennifer Jobrack (FARE) for their help in the organization of the symposium.

Bedford Park, IL, USA   Tong-Jen Fu
Bedford Park, IL, USA   Lauren S. Jackson
Bedford Park, IL, USA   Kathiravan Krishnamurthy
Madison, WI, USA   Wendy Bedale

Contents

Contributors

Timothy Adams Kellogg Company, Battle Creek, MI, USA

Wendy Bedale Food Research Institute, University of Wisconsin-Madison, Madison, WI, USA

Binaifer Bedford Division of Food Processing Science and Technology, U.S. Food and Drug Administration, Bedford Park, IL, USA

Katherine Brandt Extension, University of Minnesota, St. Paul, MN, USA

David Crownover ServSafe®, National Restaurant Association, Washington, DC, USA

Microbac Laboratories, Pittsburgh, PA, USA

Nicole Delaney Institutional Corporate Accounts, Technical Service, Ecolab, Eagan, MN, USA

Suzanne Driessen Extension, University of Minnesota, St. Paul, MN, USA

Miriam Eisenberg EcoSure, a Division of Ecolab, Naperville, IL, USA

Susan Estes Global Food Safety, PepsiCo, Inc., Barrington, IL, USA

Tong-Jen Fu Division of Food Processing Science and Technology, U.S. Food and Drug Administration, Bedford Park, IL, USA

Steven M. Gendel Division of Food Allergens, IEH Laboratories and Consulting Group, Lake Forest Park, WA, USA

Ruchi S. Gupta Northwestern University Feinberg School of Medicine, Chicago, IL, USA

Lindsay Haas Michigan Dining, University of Michigan, Ann Arbor, MI, USA

Lauren S. Jackson Division of Food Processing Science and Technology, U.S. Food and Drug Administration, Bedford Park, IL, USA

Jennifer Jobrack Food Allergy Research & Education (FARE), Skokie, IL, USA

Hal King Public Health Innovations, LLC, Fayetteville, GA, USA

Kathiravan Krishnamurthy Department of Food Science and Nutrition, Institute for Food Safety and Health, Illinois Institute of Technology, Bedford Park, IL, USA

Marissa Mafteiu School of Public Health, University of Michigan, Ann Arbor, MI, USA

Alexander M. Mitts Northwestern University Feinberg School of Medicine, Chicago, IL, USA

Alana Otto Northwestern University Feinberg School of Medicine, Chicago, IL, USA

Ann & Robert H. Lurie Children's Hospital of Chicago, Chicago, IL, USA

Madeline M. Walkner Ann & Robert H. Lurie Children's Hospital of Chicago, Chicago, IL, USA

Kathryn Whiteside Michigan Dining, University of Michigan, Ann Arbor, MI, USA

Chapter 1
Best Practices for Assessing, Managing, and Communicating Food Allergen Risks: An Introduction

Tong-Jen Fu, Lauren S. Jackson, and Kathiravan Krishnamurthy

1.1 Introduction

Food allergy is defined as an adverse health effect arising from a specific immune response that occurs reproducibly on exposure to a given food (Boyce et al. 2010). Allergic reactions to food can affect various parts of the body causing gastrointestinal discomfort (e.g., nausea, vomiting, abdominal pain, diarrhea), skin irritations (e.g., hives, atopic dermatitis, swelling of the face), respiratory disorders (e.g., wheezing, nasal congestion, breathing difficulty), or cardiovascular problems (e.g., irregular pulse, severe drop in blood pressure). Symptoms can range from mild to severe, including potentially life-threatening anaphylaxis (FDA 2017). If not treated immediately, anaphylactic reactions can be fatal. In the event of serious allergic reactions, immediate treatment with epinephrine is critical. Delayed administration of epinephrine has been associated with fatalities (Bock et al. 2001, 2007; Muñoz-Furlong and Weiss 2009).

Food allergy is a significant public health concern, affecting 5% of young children and 4% of teens and adults in the U.S. (Boyce et al. 2010). There are indications that the prevalence of food allergies is rising. Among U.S. children aged 0–17 years, the prevalence of food allergies increased by approximately 50% from 1997 to 2011 (Jackson et al. 2013). The economic impact associated with food allergies is significant. In the U.S., the total cost of caring for children with food allergies, including direct medical expenses and costs borne by the family, is estimated to be $24.8 billion/year (Gupta et al. 2013).

T.-J. Fu (✉) • L.S. Jackson
Division of Food Processing Science and Technology, U.S. Food and Drug Administration, Bedford Park, IL 60501, USA
e-mail: tongjen.fu@fda.hhs.gov

K. Krishnamurthy
Department of Food Science and Nutrition, Institute for Food Safety and Health, Illinois Institute of Technology, Bedford Park, IL 60501, USA

© Springer International Publishing AG 2018
T.-J. Fu et al. (eds.), *Food Allergens*, Food Microbiology and Food Safety,
DOI 10.1007/978-3-319-66586-3_1

Food allergens are the components within foods that trigger adverse immunologic reactions (Boyce et al. 2010; NASEM 2016). More than 160 foods have been reported to cause allergic reactions (Hefle et al. 1996). However, eight foods (milk, eggs, fish, crustacean shellfish, tree nuts, peanuts, wheat, and soybeans) account for 90% of all food allergy cases in the U.S. (U.S. Code 2004). Milk and egg allergies are more prevalent among young children, whereas seafood (fish and shellfish) allergies are more prevalent among adults (Boyce et al. 2010). Most children outgrow their allergies to milk and egg, but allergies to peanuts or tree nuts tend to persist throughout life (Sampson 2004).

To date, there is no known cure for food allergies. Strict avoidance of the allergenic food of concern remains the only effective means to prevent allergic reactions. Avoidance of allergens of concern can be difficult to achieve as foods consumed by allergic individuals are often produced, prepared or served by others. Consumers can be exposed to food allergens through a number of different routes, including packaged foods purchased at grocery/retail stores, meals consumed at restaurants or other foodservice establishments, or foods prepared at school, at community events, or at home. Inadvertent exposures can occur and result in serious consequences. Bock and Atkins (1989) reported that 50% and 75% of peanut allergic children aged 2–14 years had adverse reactions caused by accidental peanut exposure 1 year and 5 years prior to the study, respectively. A recent Canadian study of 1941 peanut allergic children yielded a 12.4% annual rate of accidental exposure to peanuts (Cherkaoui et al. 2015).

Examination of the 63 fatalities reported to the Food Allergy and Anaphylaxis Network registry from 1994 to 2006 revealed several major causes contributing to the fatal reactions (Muñoz-Furlong and Weiss 2009; Bock et al. 2001, 2007). Children or young adults aged 11–30 years comprised three-fourths of the fatalities. Eighty-seven percent of the fatalities were caused by peanuts or tree nuts. Almost half (46%) of the food items involved originated from restaurants or other food establishments. Packaged food items caused 27% of the fatalities. About 16% of the fatalities occurred in a school, child care, or university setting. Other locations where fatal reactions occurred included home and community settings such as camps, carnival, at work, or in the office.

Analysis of the anaphylaxis data obtained in the U.K. over a 21-year period (1992–2012) showed a similar trend (Turner et al. 2014). The mean age of the 124 fatalities likely to be caused by ingestion of a food allergen was 25 years. For the 95 fatalities where the food trigger was known, 73% were caused by peanuts or tree nuts. Twenty-seven percent of the fatalities were triggered by allergen ingestion in the patient's own home, whereas 20% occurred in restaurants. In the 100 cases where the food source was identified, 27% were caused by consumption of prepackaged foods, while 59% of reactions were due to food products provided by a catering establishment.

Allergens can be introduced into food via several routes. They can be added intentionally as ingredients. They can be inadvertently incorporated in packaged foods due to incorrect labeling, production errors or allergen cross-contact (Jackson et al. 2008). Foods served in restaurants or other food establishments and in the home can inadvertently become contaminated with allergens through similar mechanisms. Finally, allergens can be potentially introduced into food through modern biotechnology methods (Nordlee et al. 1996).

From a public health perspective, the key principle in managing food allergen risk is to prevent inadvertent exposure to sensitive individuals. This can be achieved by requiring accurate food allergen labeling on packaged food, by clearly communicating the presence of allergenic ingredients in meals served in restaurants and other food establishments, and by placing appropriate control measures to prevent allergen cross-contact during food manufacture, storage, preparation, and service. In the case of genetically engineered foods, risk management can be accomplished by conducting a premarket assessment and evaluating the allergenic potential of newly introduced proteins (FAO/WHO 2001; Codex 2003).

In the past two decades, significant advances have been made in our understanding of food allergens, in the identification of risk factors associated with them, and in the development of control measures to mitigate these risks. Significant progress has also been made in our ability to manage food allergies as a society, including improvements in the ability to recognize food allergy symptoms and a greater availability of epinephrine in schools and other public entities. Key advances and challenges that remain on allergen control in food manufacturing facilities, foodservice establishments, other community settings, and in the home are summarized below.

1.2   Managing Allergens in Food: Current Status

1.2.1   Allergen Management in the Packaged Food Industry

Consumers rely on accurate food labels to reveal the presence of allergenic ingredients. The Food Allergen Labeling and Consumer Protection Act (FALCPA) requires manufacturers of packaged food to clearly disclose ingredients derived from eight major allergenic food groups (U.S. Code 2004). Similar food allergen labeling regulations have also been enacted in other countries, although the list of priority allergens, the process for modifying this list, and the approach to labeling declaration may differ (Gendel 2012).

FALCPA has been in effect for more than a decade, but undeclared allergens continue to be the leading cause of food recalls in the U.S. (Gendel and Zhu 2013). The number of allergen recalls for FDA-regulated food products more than doubled (from 78 to 189) during fiscal years 2007 to 2012 (Gendel and Zhu 2013). The percentage of reportable food incidences due to undeclared allergens increased from 30% in fiscal year 2010 to 47% in fiscal year 2014 (FDA 2016a). Bakery products were the most frequently recalled food type, and milk was the most frequently undeclared major food allergen, possibly due to the many forms of milk and milk-derived ingredients that are used in packaged foods (Gendel and Zhu 2013).

Root-cause analyses suggested that failures in label controls were the leading cause of food allergen recalls, responsible for 67% of all recalls with a known root cause (Gendel and Zhu 2013). Failure to declare major food allergens, use of the wrong package or wrong label, terminology problems, or lack of carry-through of an allergen declaration from an ingredient were the major causes of labeling errors. An analysis of recalls on products regulated by the U.S. Food Safety and Inspection Service (meat, poultry, and certain egg products) showed similar results (Hale 2017). In 263 recalls that occurred during 2005–2015, labeling errors (incorrect label, package, or ingredient statement, missing ingredient statement, or statement omitting allergens known to be present) were the cause for the majority (63%) of the recalls. These results highlight the need for the packaged food industry to research, develop, and implement effective label management programs to ensure accurate and proper allergen labeling.

Current labeling regulations deal only with allergens that are knowingly added as ingredients. They do not address allergens inadvertently introduced into food as a result of production errors or cross-contact during manufacture and storage. Allergen cross-contact was the cause of 8% of the FDA and 10% of the FSIS recalls mentioned above (Gendel and Zhu 2013; Hale 2017). Allergen cross-contact that may arise in food manufacturing facilities include in-process or post-process cross-contact, errors in handling of rework, improper production sequences which result in one product contaminating a subsequent product, and insufficient or ineffective equipment cleaning/sanitation procedures at product changeover (Jackson et al. 2008). To prevent the inadvertent introduction of allergens into food, it is critical for manufacturers to develop and implement preventive control measures through the application of supplier controls, production sequencing, physical separation, cleaning and sanitation, and process design (Gendel and Zhu 2013).

The importance of implementing food allergen control programs has been recognized by the U.S. government, as reflected in the Food Safety Modernization Act (FSMA) that was signed into law in 2011. The FSMA Preventive Controls for Human Food Rule (FDA 2016b) requires each food production facility to prepare and implement a written food safety plan describing practices and procedures for controlling food safety hazards. Required in the food safety plan are allergen preventive controls, which must be implemented to provide assurance that allergen hazards are significantly minimized or prevented at all phases of food manufacture and storage.

The Preventive Controls Rule does not specify what considerations need to be included in an allergen control plan (ACP), but elements of effective allergen controls have been described (Taylor and Hefle 2005). The key principles and considerations involved in successful allergen management during food production have also been discussed (Stone and Yeung 2010). A number of other resources aimed at helping food manufacturers develop their own allergen control programs include "Managing Allergens in Food Processing Establishments," published by the Grocery Manufacturers Association (GMA 2009), and "Components of an effective allergen control plan: A framework for food processors," developed by the University of Nebraska Food Allergy Research and Resource Program (FARRP 2009).

Despite the availability of resources, broad implementation of allergen controls across the packaged food industry is still lacking. A recent survey of food allergen control practices in the U.S. indicated that awareness and the use of allergen controls have increased significantly in the last decade, but important gaps still exist (Gendel et al. 2013). Approximately 20% of surveyed facilities did not declare all of the major food allergens used, and approximately 30% of these facilities did not protect products from cross-contact during production. A relatively large proportion of small food manufacturing facilities did not implement all of the controls needed to prevent cross-contact. While written procedures and records were used by all of the large facilities, only 40% of small facilities had written cleaning procedures, and only 28% of the small facilities kept written records. The report suggested that these gaps can be addressed through development of guidance and education programs (Gendel et al. 2013).

Small food manufacturing facilities are in need of allergen control programs that are tailored to their specific facilities, processing lines, equipment, operations, and training needs. Availability of model ACP and clear demonstration (e.g., through outreach and other education programs) of best practices for each stage of food manufacture including hazard analysis, supplier controls, allergen label controls, cross-contact controls, and staff training are essential to enable small facilities to implement and effectively execute allergen control programs.

Even with the advances made in allergen removal from shared equipment, tools, or processing lines (see Chaps. 8 and 10), allergen cross-contact can still occur due to the use of improperly designed equipment or less effective cleaning methods, e.g., dry-cleaning instead of the more effective wet-cleaning methods in chocolate manufacturing facilities (Jackson et al. 2008). Precautionary allergen labeling (PAL) or advisory labeling such as "may contain" or "manufactured on equipment that also processes" is used by food manufacturers on a voluntary basis to alert consumers to the possible presence of an allergen in a food product. FDA advises that advisory labeling should be truthful and not misleading and should not be used in lieu of adherence to current Good Manufacturing Practices (FDA 2006).

The use of PAL in packaged foods is widespread. A survey of more than 20,000 unique products sold in U.S. supermarkets found that 17% contained advisory labels, and more than 50% of the chocolate candy or cookie products surveyed had PAL (Pieretti et al. 2009). Twenty-five different types of advisory statements were used among 98 products that contained advisory labels (Pieretti et al. 2009). The effectiveness of PAL, as currently practiced, in alerting consumers about food allergen risks has been questioned (NASEM 2016; DunnGalvin et al. 2015; Marchisotto et al. 2017). Multiple studies analyzing allergen content in food products have demonstrated that the presence or absence of PAL does not correlate with health risks. Food products with PAL often do not contain detectable allergen residues, while products without PAL can contain significant levels of allergens (Crotty and Taylor 2010; Ford et al. 2010; Khuda et al. 2016; Bedford et al. 2017). Similarly, no correlation exists between the level of detectable allergens and the type of statement used in PAL (Crotty and Taylor 2010; Bedford et al. 2017). The lack of correlation between PAL and allergen risks has led to the recommendation that allergic consumers should avoid advisory-labeled products (Taylor and Hefle 2006; Ford et al. 2010).

Inconsistency in the use of PAL in packaged food has generated a great deal of confusion and misconceptions among consumers. It has unnecessarily limited the food choice for many allergic individuals and has also contributed to reduced avoidance and increased risk-taking by allergic consumers who often ignore PAL (Hefle et al. 2007; DunnGalvin et al. 2015; Marchisotto et al. 2017). A survey of consumers' knowledge of PAL and its impact on purchasing habits of food-allergic individuals and caregivers in the U.S. and Canada reported that up to 40% of surveyed consumers purchased products with PAL and that buying practices varied depending on the particular PAL used. More than a third of consumers perceived that different PAL statements indicated different amounts of allergens (Marchisotto et al. 2017).

There is a need for standardization of PAL to help consumers make informed food choices (Marchisotto et al. 2017). Stakeholders agree that PAL should be risk-based, should indicate the potential unintended presence of an allergen at or above a reference dose, and should be consistent across food products. The decision-making criteria should be transparent and clearly communicated to all stakeholders (DunnGalvin et al. 2015). Currently, no agreed-upon reference doses exist. Extensive work is underway using quantitative risk assessment approaches to determine individual threshold doses, which will inform the development of reference doses (NASEM 2016). Another key challenge in achieving a successful PAL program is the limitations surrounding available analytical methods for detecting allergens in food. Many studies have shown that different allergen test kits may return different analytical results for the same food sample (Khuda et al. 2016; Bedford et al. 2017). Various factors may contribute to the inconsistency in allergen quantitation among different test kits (Fu and Maks 2013), but this issue will need to be resolved to correctly quantify allergen residues in foods relative to target reference doses.

1.2.2   Allergen Management in the Foodservice Industry

It is estimated that American households spend nearly 44% of their total food expenditures on food prepared outside the home (ERS 2017). The U.S. restaurant industry serves more than 130 million customers each day (NRA 2014) and employs 14.7 million people across roughly one million locations (NRA 2017).

Dining outside the home poses a significant challenge for allergic consumers. Inadvertent exposure to food allergens in restaurants is not uncommon. In a recent study, 21.8% of the 110 restaurant personnel surveyed indicated that food-related allergic reactions had occurred at their restaurants in the past 12 months (Lee and Xu 2015). Among the 63 fatal food anaphylactic reactions occurred in 1994–2006, 29 cases (46%) were caused by food served in, or provided by, restaurants or other food establishments, including sit-down restaurants, fast food establishments, university dining halls, school cafeteria, ice cream shops, and catered functions (Weiss and Muñoz-Furlong 2008).

A study of 156 peanut and tree nut allergic reactions caused by foods purchased at restaurants and other food establishments found that Asian food restaurants (19%), ice cream shops (14%), and bakeries/doughnut shops (13%) were the locations most commonly involved. Among meal items, desserts (cakes and ice cream) were the most common cause of the reactions (43%). Only 45% of the 106 participants with previously diagnosed food allergies gave prior notification about their allergies to the establishment. In 78% of these incidents, someone in the establishment knew that the food contained peanut or tree nut, but the presence of these ingredients was not disclosed. In 50% of these incidents, the allergens of concern were hidden in sauces, dressings, egg rolls, etc. Twenty-two percent of the exposures were from contamination caused primarily by shared cooking/serving supplies (Furlong et al. 2001).

Clearly, failure of food allergic individuals to make their allergies known, lack of understanding of the serious nature of food allergy, inability of service personnel to clearly communicate allergen information, and absence of proper food preparation procedures to avoid allergen cross-contact all contributed to the inadvertent exposure of allergic consumers to the harmful allergens. To prevent accidental exposure and tragic incidents, foodservice operations will need to increase awareness of food allergy and allergens, better understand allergen risk factors, develop tools and programs that foster clear communication of allergen information to customers and among staff, and install controls that prevent allergen cross-contact. Effective training programs are also needed to educate staff on allergen control best practices.

A number of surveys of restaurant personnel have pointed to the significant knowledge gaps that exist in the foodservice industry with respect to food allergy awareness and allergen controls. A survey of 100 food handlers (manager, servers, and chefs) in 100 food establishments in New York that offered different cuisines and types of service (full service, fast food, and takeout) found that while the majority (72%) of surveyed personnel expressed a high comfort level in providing safe meals to allergic consumers, there were deficits in their knowledge of food allergies. For example, 24% of the respondents did not recognize that trace amounts of a food can trigger a reaction, and 54% did not recognize common means of cross-contact of foods (Ahuja and Sicherer 2007).

The CDC Environmental Health Specialists Network (EHS-Net) surveyed personnel at 278 restaurants located in six sites during 2014–2015 and reported that only 44.4% of restaurant managers, 40.8% of food workers, and 33.3% of servers interviewed reported receiving food allergy training (Radke et al. 2017). In an online survey to assess the food allergy knowledge, attitudes, and preparedness of restaurant managerial staff, Lee and Xu (2015) showed that although 80% of the 110 participants had been trained about food allergies and approximately 70% of participants provided food allergy training to their employees, the food allergy knowledge varied greatly among the participants. Forty percent of the participants were not able to point out that soy and fish are among the major allergens. More than 80% of the participants were not aware of FALCPA food allergen labeling requirements. In a survey of 16 managers from full-service restaurants that accommodate allergen-free orders upon request, it was found that managers of chain restaurants or independent restaurants located in chain hotels were more likely to have received training and to have access to educational materials than managers of independent restaurants. However, training on allergen risk communication between the front-of-house and back-of-house staff or between restaurant staff and customers with food allergies is lacking (Wen and Kwon 2016). These survey findings highlight the need for appropriate and adequate training for restaurant operators and the need for risk-based and structured training materials that restaurants can use to train their staff.

The importance of increasing awareness of food allergy and food allergens and ensuring proper training of foodservice personnel has also been recognized by regulatory agencies. In the U.S., restaurants and other food establishments are regulated by the state and local government agencies. The FDA works with these agencies through the Conference for Food Protection and through development of a model Food Code (Gendel 2014). The Food Code provides practical, science-based guidance and enforceable provisions for preventing risks of foodborne illness and serves as a model code for adoption by local, state, and other jurisdictions to apply to all food establishments that sell food directly to consumers. In the past decade, the FDA Food Code has been revised several times to include food allergen information and procedures for controlling allergens. The 2005 revision of the Food Code contained a definition of the "major food allergens," specifying that the person in charge of a food establishment shall have an understanding of foods identified as major food allergens and of the symptoms resulting from exposure to food allergens in sensitive individuals. Other added information included an integration of FALCPA's labeling provisions to food packaged at the retail level. The 2005 Food Code also recommended the use of a rigorous sanitation regimen to prevent cross-contact between allergenic and non-allergenic ingredients (FDA 2015). Additions in the 2009 revision of the Food Code included the requirement that employees are properly trained in food allergy awareness as it relates to their assigned duties (FDA 2015). The majority of U.S. states have adopted the recent versions of Food Code with updated allergen requirements (see Chap. 3).

In the U.S., meals served in restaurants or other food establishments, except those packaged for direct sale in states that have adopted the updated FDA Food Codes, are not subject to the FACLPA food allergen labeling requirements. Allergic consumers rely on servers to accurately disclose allergen information in menu items. However, the majority (82%) of 110 restaurant managers surveyed considered it is the customers' responsibility to inform the servers of their special needs. About 38% of the respondents either disagreed or strongly disagreed that restaurants should be responsible for asking customers about their food allergy needs (Lee and Xu 2015). This finding suggests that in the event of miscommunication or absence of communication, accidental exposure could occur. To address this issue, some states have enacted laws that require food establishments to include menu notice for customers to inform servers of their food allergies (see Chap. 3). It is worth mentioning that in Europe, the new food information regulation 1169/2011 added a provision requiring allergen information be provided for non-prepackaged food sold in the European Union, including foods sold in retail, in catering, and food prepacked on the premises for direct sale (EU 2011; Hattersley and King 2014). The regulation allows each member state to establish rules regarding how the allergen information is presented, e.g., on signs, menus, or communicated through the server when requested (Leitch and McIntosh 2014).

Despite the inclusion of food allergy awareness, employee training, and allergen cleaning and sanitation in the updated Food Codes, there is currently a lack of national guidelines applicable to the foodservice industry for assessing, managing, and communicating food allergen risks in restaurants and other food establishments. Limited information has been provided by consumer advocacy and trade groups. The booklet "Welcoming Guests with Food Allergies" developed by Food Allergy and Anaphylaxis Network (FAAN, now part of the Food Allergy Research and Education, FARE) (FAAN 2010) provides general information about food allergies and suggests strategies for managers, front-of-house staff, and back-of-house staff to ensure a safe dining experience for guests with food allergies. Considerations for creating a written plan for handling guests with food allergies, for staff training, and for responding to an allergy emergency were also included. The National Restaurant Association (NRA) has developed educational materials and programs aimed at educating restaurant and foodservice workers on aspects of allergen control. The ServSafe allergens online course covers such topics as defining food allergies, identifying allergens, communicating with guests, preventing cross-contact, proper food preparation and cleaning methods, food labels, and dealing with emergencies, etc. (NRA 2017). This course is an approved training resource to meet the allergen training requirements in Michigan and Rhode Island (see Chap. 9).

Even with increased efforts at the federal and state levels to include food allergen controls as preventative measures against foodborne illness, there remains a lack of widespread adoption of proper allergen control measures, particularly in small and standalone foodservice operations. In the EHS-Net survey, it was found that only 55.2% of managers reported that their restaurants had ingredient lists for menu items while 25.3% reported having no lists. The majority of restaurant managers (78.0%) reported that dedicated utensils or equipment were not used in their restaurants for preparing allergen-free food. Only 7.6% of restaurants reserved an area in the kitchen for preparing allergen-free food, and only 10.1% had a dedicated fryer for cooking allergen-free food (Radke et al. 2017).

The foodservice industry comprises many different operations (full service, fast food, take out, etc.) that serve a wide range of meal options. These operations, especially the small and standalone operations, will need guidance and a clear step-by-step demonstration of how to incorporate available allergen control best practices into their own allergen management plans, including implementation of ingredient/menu controls, ingredient/food storage and preparation controls, cleaning/sanitation controls, and education of managers and all staff involved. To achieve broad adoption of best practices, relevant education and outreach programs and effective dissemination of available allergen management information and tools to all types of operations are needed.

1.2.3 Allergen Management at Community Settings and at Home

Inadvertent exposure to food allergens can occur anywhere food is served and consumed, including the allergic individual's own home. In one study, 37% of children with peanut allergy were accidentally exposed to peanuts at home in the preceding year (Cherkaoui et al. 2015). Among the 63 fatalities in 1994–2006 caused by anaphylactic reactions to food, 13 (20%) occurred at the person's own home (Bock et al. 2001, 2007). Twenty-seven percent of the fatalities in the U.K. study were triggered by allergen ingestion in the individual's home (Turner et al. 2014). Consumer groups such as the Food Allergy Research and Education (FARE) have provided guidance to assist allergic individuals and their caregivers in preventing inadvertent ingestion of food allergens at home. Considerations such as whether to ban the allergens from entering the home, reading labels, proper food preparation to avoid cross-contact, segregation of allergens during storage, and how to eliminate cross-contact risks in the kitchen are provided (FARE 2017a). Such information has proven useful for parents of allergic children to develop and implement food allergen management plans in their homes (see Chap. 13).

Schools, colleges, and other educational settings are also places where food allergic reactions can occur. Greater than 15% of children with food allergies had reactions in school and day care (Nowak-Wegrzyn et al. 2001; Sicherer et al. 2001a, 2001b). The reactions were primarily caused by exposure to food containing milk, peanut, and egg (Nowak-Wegrzyn et al. 2001). To reduce allergic reactions and to improve the ability of U.S. schools to respond to emergency food allergic reactions, the CDC in 2013 released the first national guidelines for school food allergy management. The "Voluntary Guidelines for Managing Food Allergies in Schools and Early Care and Education Programs" called for the development of food allergy management and prevention plans that address five priority areas: (1) ensuring daily management of food allergies for individual children, (2) preparing for food allergy emergencies, (3) providing professional development on food allergies for staff, (4) educating children and family members about food allergies, and (5) creating and maintaining a healthy and safe educational environment (CDC 2013).

FARE has worked with schools, universities, and other community groups to develop educational programs and training tools for a wide range of audiences. The document "School Guidelines for Managing Students with Food Allergies" outlines responsibilities for the family, school, and students to minimize risks and provide a safe environment for food-allergic students. It recommends that the family notify the school of the child's allergy, work with the school to develop a plan that accommodates the child's needs, and educate the child in self-management of food allergy. The schools need to ensure that all staff who interact with the student understand food allergy, recognize symptoms, know what to do in an emergency, and eliminate the use of food allergens in the allergic student's meals, educational tools, and other school activities. The students, based on their developmental level, should be proactive in the care and management of their food allergies and reactions (FARE 2017b).

These guidelines, however, do not offer recommendations for foodservice staff on measures to prevent inadvertent exposure of allergic students to allergens. To fill this gap, the Institute of Child Nutrition has developed a training program (Managing Food Allergies in School Nutrition Programs) for school nutrition professionals to learn how to accommodate students with food allergies. The program includes sections on general food allergy principles, reading and managing food labels, accommodating students with food-related disabilities, avoiding cross-contact, and promoting food allergy management in schools (ICN 2014).

A greater number of fatalities caused by anaphylactic reactions to food occurred in teens and young adults than in other age groups (Muñoz-Furlong and Weiss 2009; Turner et al. 2014). College-aged subjects (18–22 years) accounted for 25% of the 63 fatal allergic reactions reported in 1994–2006, and 50% of these fatalities occurred on a college campus (Bock et al. 2001, 2007). Allergic individuals attending college need additional assistance as this may be the first time these individuals are away from home and from the care of their parents. Teens and young adults with food allergy often exhibit risk-taking behaviors. In a survey of 174 food-allergic teens, 54% indicated they had purposefully ingested a potentially unsafe food (Sampson et al. 2006). In 2015, FARE published the "Pilot Guidelines for Managing Food Allergies in Higher Education" which is a comprehensive best practice guide that helps colleges and universities effectively manage food allergy (FARE 2017c). Topics covered included food allergy management on campus, food allergy policy, emergency response plans and training. The guidelines also provided recommendations for college dining services regarding training, label reading, back-of-house and front-of-house policies, and student responsibilities, etc. These guidelines have been adopted by universities to develop food allergen control programs (see Chap. 14).

It is estimated that each year, more than 11 million children and adults attend a camp (e.g., day camps, residential camps, sports camps, or travel camps). A review of health records from 170 summer camps across the U.S. and Canada indicated that 2.5% of the 122,424 campers had documented food allergies. Peanut/tree nut (81%), seafood (17.4%), and egg (8.5%) were the top three allergies reported (Schellpfeffer et al. 2017). To ensure the safety of children with food allergy, it is important that camps have established food allergy policies. FARE has developed guidelines for camps and recommended steps for parents, camp staff, and campers themselves to minimize the risk of accidental exposure to a food allergen. The guidelines also recommended having medications and procedures in place to deal with accidental ingestion (FARE 2017d).

The majority of available guidelines developed to protect allergic individuals in community settings focus primarily on management of food allergy reactions in the event of accidental exposure. Significant gaps still exist in the development and implementation of control measures to prevent accidental exposures in these settings where food is prepared and served. Many questions remain to be addressed, including: are there additional risk factors that need to be understood, how can food allergen information be shared with foodservice operators, how can allergen control best practices (ingredient control, accurate communication of allergen information, cross-contact control, allergen cleaning, etc.) be implemented, what training should event staff, temporary workers, and volunteers undergo before preparing and serving foods, and how can it be ensured that proper risk communication and preventive measures are implemented and practiced? There is a need for easy-to-understand, step-by-step guidance on best practices to control food allergens during the preparation and serving of food in schools, day cares, camps, and other community settings. Many of these gaps may be addressed by adopting the best practices identified and implemented in the packaged food and foodservice industries.

1.3   Managing Food Allergy: Emergency Treatment of Anaphylaxis

Despite increased awareness and efforts to implement allergen control measures across many stages of the food chain, allergic individuals may still experience accidental exposure (Bock and Atkins 1989; Sicherer et al. 2001a, 2001b; Fleischer et al. 2012; Cherkaoui et al. 2015). Food-allergic individuals must always be prepared to respond to potentially severe reactions in the event of accidental ingestion of an allergen. Equally critical is the ability of people who serve food and who are responsible for the well-being of allergenic individuals to recognize food allergy symptoms. Available guidelines discussed above all stress the importance of training foodservice personnel, caregivers, school staff, or camp staff to quickly recognize food allergy symptoms and to effectively respond to a food allergy emergency.

Immediate use of epinephrine can save lives. About 70% of the 63 fatalities due to anaphylactic reactions to food did not receive epinephrine in a timely manner (Muñoz-Furlong and Weiss 2009). In a study conducted in the U.K., epinephrine was used correctly and in a timely manner by just 19% of the victims (Turner et al. 2014). Only 29.9% of children aged 15 months to 3 years who experienced severe allergic reactions were treated with epinephrine (Fleischer et al. 2012). Factors resulting in undertreatment included lack of recognition of severity, unavailability of epinephrine, and fears about epinephrine administration (Fleischer et al. 2012).

Significant efforts have been made to increase the availability of epinephrine in schools. Almost every state in the U.S. has passed legislation regarding stocking undesignated epinephrine auto-injectors in K-12 schools (FARE 2017e); however, gaps still exist with respect to training of school staff in recognition of symptoms and in administration of epinephrine to treat anaphylaxis (White et al. 2016). In a study examining the use of epinephrine auto-injectors in Chicago Public Schools during the 2012–2013 school year, DeSantiago-Cardenas et al. (2015) found that more than half (55.0%) of all district-issued epinephrine auto-injectors were administered for first-time anaphylactic events. These findings suggest that anaphylaxis can occur in children with no previously known allergies and highlight the importance of stocking undesignated epinephrine in schools (see Chap. 2).

Allergic reactions can occur in any location where food is consumed. For individuals who experience a reaction but do not carry epinephrine with them, access to undesignated epinephrine auto-injectors can be critical and life-saving. Consumer advocacy groups such as FARE have worked with legislative bodies at the national and state levels to increase the availability of undesignated epinephrine auto-injectors and to ensure that emergency epinephrine capabilities are in place in public venues (see Chap. 4). Currently, a number of states have passed legislations that permit public entities (e.g., camps, restaurants, amusement parks, and sports arenas) to stock undesignated epinephrine auto-injectors for emergency use and require training of staff to recognize systemic reactions to food and to properly administer epinephrine (McEnrue and Procopio 2016; ACA 2017).

1.4   Best Practices for Assessing, Managing, and Communicating Food Allergen Risks

Protecting consumers with food allergies requires vigilance from everyone involved in the manufacture, preparation, and serving of foods. A great deal of knowledge about food allergens and approaches to controlling them has been accumulated, particularly in the packaged food industry. However, detailed guidance and step-by-step demonstrations of how to implement allergen controls, especially for small food manufacturing facilities, are lacking. There is also a shortage of customized tools or model plans for hazard analysis and allergen management that address the needs of specific operations in other environments where food is prepared and served, including restaurants, community settings, or at home.

This book provides the most up-to-date information on allergen controls applicable to various stages of the food chain. The resources presented and experience shared will assist all stakeholders in establishing best practices for the assessment, management, and communication of risks associated with food allergens. Chapters 2–4 provide an overview of the public health impact of food allergy and legislative initiatives that are on-going to improve food allergy and allergen management across various sectors of the industry and in many communities. Analysis of recall data or surveys of industry practices are valuable in identifying risk factors and gaps that need to be addressed. Chapter 5 reviews the available recall databases and discusses the use of recall data analytics to support food safety hazard analysis and to identify trends and emerging issues.

Chapters 6–8 discuss allergen management best practices that may be implemented in food manufacturing facilities. Key components and considerations of an effective allergen label management program are discussed in Chap. 6. Main features of an allergen management program focusing on preventing allergen cross-contact at various stages of food production operations are covered in Chap. 7. This chapter also discusses verification and validation strategies for allergen sanitation. An in-depth discussion of allergen cleaning and sanitation best practices in a food production environment is presented in Chap. 8.

Chapter 9 discusses findings from a survey conducted by the National Restaurant Association to evaluate food allergen awareness and training policies in the restaurant and foodservice industry. Chapter 10 summarizes the elements of a solid food allergen management program for restaurants to ensure correct allergen information is shared and allergen cross-contact is minimized. Methods and factors for successful cleaning are also discussed. Chapter 11 highlights best practices for managing food allergens in quick-service restaurants and discusses key aspects to consider in preventing food allergic reactions throughout the operation, including controlling the supply chain and ingredients, managing restaurant operations, and communicating with customers.

Best practices in developing training programs for assessing, managing, and communicating allergen risks for restaurants and other foodservice operations are presented in Chap. 12. Finally, key points to consider when developing customized allergen control plans for the home and for dining services on college campuses are discussed in Chaps. 13 and 14, respectively.

1.5   Concluding Remarks

The public health issues associated with food allergens can only be tackled by taking an integrative approach that involves all sectors of the food chain, including the packaged food industry, the foodservice industry, community groups, and consumers. Success will require broad awareness of the severity of food allergic reactions, understanding of risk factors, and accurate implementation of allergen control best practices. Effective training of relevant personnel and improved communication with consumers with food allergies are critical.

This book documents best practices in managing allergen risks at various stages of the food chain. Effective dissemination of these resources to stakeholders is needed. University extension services and state and local public health agencies have traditionally played a key role in providing education and outreach on food safety issues and controls to small food manufacturing facilities and food establishments. It is important to develop a network of experts in extension and outreach services throughout the U.S. This will enable dissemination of accurate and complete allergen management information to all sizes of food production and foodservice operations, and eventually achieve broad and successful implementation of effective allergen control programs across the entire food chain.

Regulations aimed at reducing allergic reactions caused by packaged food and in foodservice environments have been enacted to improve adoption of best practices. Enforcing compliance with these regulations will ensure that allergen controls are effectively implemented. Training of state/local inspectors to recognize and verify proper food allergen control measures is critical to the success of these regulatory efforts.

The information presented in this book provides valuable resources for stakeholders to develop structured, risk-based training programs. Consumers and other food safety professionals will also benefit from the knowledge gained, and be able to recognize and understand food allergen control measures that are put in place across the food chain.

References

ACA (American Camp Association). 2017. Epinephrine auto-injectors accessibility laws and camps. Available at: https://www.acacamps.org/resource-library/public-policy/epinephrine-auto-injectors-accessibility-laws-camps. Accessed 22 June 2017.

Ahuja, R., and S.H. Sicherer. 2007. Food allergy management from the perspective of restaurant and food establishment personnel. *Annals of Allergy, Asthma & Immunology*. 98: 344–348.

Bedford, B., Y. Yu, X. Wang, E.A.E. Garber, and L.S. Jackson. 2017. A limited survey of dark chocolate bars obtained in the United States for undeclared milk and peanut allergens. *Journal of Food Protection*. 80: 692–702.

Bock, S.A., and F.M. Atkins. 1989. The natural history of peanut allergy. *Journal of Allergy and Clinical Immunology*. 83: 900–904.

Bock, S.A., A. Munoz-Furlong, and H.A. Sampson. 2001. Fatalities due to anaphylactic reactions to foods. *Journal of Allergy and Clinical Immunology*. 107: 191–193.

Bock, S.A., A. Muñoz-Furlong, and H.A. Sampson. 2007. Further fatalities caused by anaphylactic reactions to food, 2001-2006. *Journal of Allergy and Clinical Immunology*. 119: 1016–1018.

Boyce, J.A., A. Assa'ad, W. Burks, S.M. Jones, H.A. Sampson, R.A. Wood, M. Plaut, S.F. Cooper, M.J. Fenton, S.H. Arshad, S.L. Bahna, L.A. Beck, C. Byrd-Bredbenner, C.A. Camargo Jr., L. Eichenfield, G.T. Furuta, J.M. Hanifin, C. Jones, M. Kraft, B.D. Levy, P. Lieberman, S. Luccioli, K.M. McCall, L.C. Schneider, R.A. Simon, F.E. Simons, S.J. Teach, B.P. Yawn, and J.M. Schwaninger. 2010. Guidelines for the diagnosis and management of food allergy in the United States: Report of the NIAID-sponsored expert panel. *Journal of Allergy and Clinical Immunology*. 126: S1–58.

CDC (Centers for Disease Control and Prevention). 2013. Voluntary guidelines for managing food allergies in schools and early care and education programs. Available at: https://www.cdc.gov/healthyyouth/foodallergies/pdf/13_243135_a_food_allergy_web_508.pdf. Accessed 20 March 2017.

Cherkaoui, S., M. Ben-Shoshan, R. Alizadehfar, Y. Asai, E. Chan, S. Cheuk, G. Shand, Y. St-Pierre, L. Harada, M. Allen, and A. Clarke. 2015. Accidental exposures to peanut in a large cohort of Canadian children with peanut allergy. *Clinical and Translational Allergy*. 5: 16.

Codex Alimentarious Commission. 2003. Guideline for the conduct of food safety assessment of foods derived from recombinant-DNA plants. Annex 1: Assessment of possible allergenicity. Available at: www.fao.org/input/download/standards/10021/CXG_045e.pdf. Accessed 20 March 2017.

Crotty, M.P., and S.L. Taylor. 2010. Risks associated with foods having advisory milk labeling. *Journal of Allergy and Clinical Immunology*. 125: 935–937.

DeSantiago-Cardenas, L., V.R. Rivkina, S.A. Whyte, B.C. Harvey-Gintoft, B.J. Bunning, and R.S. Gupta. 2015. Emergency epinephrine use for food allergy reactions in Chicago Public Schools. *American Journal of Preventive Medicine*. 48: 170–173.

DunnGalvin, A., C.H. Chan, R. Crevel, K. Grimshaw, R. Poms, S. Schnadt, S.L. Taylor, P. Turner, K.J. Allen, M. Austin, A. Baka, J.L. Baumert, S. Baumgartner, K. Beyer, L. Bucchini, M. Fernández-Rivas, K. Grinter, G.F. Houben, J. Hourihane, F. Kenna, A.G. Kruizinga, G. Lack, C.B. Madsen, E.N.C. Mills, N.G. Papadopoulos, A. Alldrick, L. Regent, R. Sherlock, J.M. Wal, and G. Roberts. 2015. Precautionary allergen labelling: perspectives from key stakeholder groups. *Allergy*. 70: 1039–1051.

ERS (Economic Research Service). 2017. Food expenditure. Available at: https://www.ers.usda.
    gov/data-products/food-expenditures/food-expenditures/#Food%20Expenditures. Accessed 20
    March 2017.
EU (European Union). 2011. Regulation (EU) No 1169/2011 of the European Parliament and of
    the council of 25 October 2011 on the provision of food information to consumers. *Official
    Journal of the European Union.* L304: 18–63.
FAAN (Food Allergy & Anaphylaxis Network). 2010. Welcoming guests with food allergies.
    Available at: https://www.foodallergy.org/file/welcoming-guests-faan.pdf. Accessed 20 March
    2017.
FAO/WHO (Food and Agricultural Organization/Word Health Organization). 2001. Evaluation of
    allergenicity of genetically modified foods. Report of a joint FAO/WHO expert consultation on
    allergenicity of foods derived from biotechnology. January 22–25, Rome, Italy.
FARE (Food Allergy Research and Education). 2017a. Managing food allergies at home. Available
    at: https://www.foodallergy.org/managing-food-allergies/at-home. Accessed 20 March 2017.
———. 2017b. School guidelines for managing students with food allergies. Available at: https://
    www.foodallergy.org/file/school-guidelines-faan.pdf. Accessed 20 March 2017.
———. 2017c. Pilot guidelines for managing food allergies in higher education. Available at:
    https://www.foodallergy.org/file/college-pilot-guidelines.pdf. Accessed 20 March 2017.
———. 2017d. Managing food allergies at camp. Available at: https://www.foodallergy.org/
    managing-food-allergies/at-camp. Accessed 20 March 2017.
———. 2017e. Epinephrine at school. Available at: https://www.foodallergy.org/advocacy/
    epinephrine-at-school. Accessed 20 March 2017.
FARRP (Food Allergy Research and Resource Program). 2009. Components of an effective allergen
    control plan: A framework for food processors. Available at: http://farrp.unl.edu/3fcc9e7c-
    9430-4988-99a0-96248e5a28f7.pdf. Accessed 20 March 2017.
FDA (Food and Drug Administration). 2006. Guidance for industry: Questions and answers
    regarding food allergens, including the Food Allergen Labeling and Consumer Protection Act
    of 2004. Available at: https://www.fda.gov/food/guidanceregulation/guidancedocumentsregu-
    latoryinformation/allergens/ucm059116.htm. Accessed 30 April 2017.
———. 2015. FDA Food Code. http://www.fda.gov/Food/GuidanceRegulation/RetailFoodProtection/
    FoodCode/default.htm. Accessed 2 December 2016.
———. 2016a. The reportable food registry: A five year overview of targeting inspection resources
    and identifying patterns of adulteration, September 8, 2009–September 7, 2014. Available
    at: https://www.fda.gov/downloads/Food/ComplianceEnforcement/RFR/UCM502117.pdf.
    Accessed 20 March 2017.
———. 2016b. Current good manufacturing practice, hazard analysis, and risk-based preventive
    controls for human food. Available at: https://www.fda.gov/food/guidanceregulation/fsma/
    ucm334115.htm. Accessed 2 December 2016.
———. 2017. Food allergies: What you need to know. Available at: https://www.fda.gov/down-
    loads/Food/ResourcesForYou/Consumers/UCM220117.pdf. Accessed 20 April 2017.
Fleischer, D.M., T.T. Perry, D. Atkins, R.A. Wood, A.W. Burks, S.M. Jones, A.K. Henning,
    D. Stablein, H.A. Sampson, and S.H. Sicherer. 2012. Allergic reactions to foods in preschool-
    aged children in a prospective observational food allergy study. *Pediatrics.* 130: e25–e32.
Ford, L.S., S.L. Taylor, R. Pacenza, L.M. Niemann, D.M. Lambrecht, and S.H. Sicherer. 2010.
    Food allergen advisory labeling and product contamination with egg, milk, and peanut. *Journal
    of Allergy and Clinical Immunology.* 126: 384–385.
Fu, T.J., and N. Maks. 2013. Impact of thermal processing on ELISA detection of peanut allergens.
    Journal of Agricultural and Food Chemistry. 61: 5649–5658.
Furlong, T.J., J. DeSimone, and S.H. Sicherer. 2001. Peanut and tree nut allergic reactions in
    restaurants and other food establishments. *Journal of Allergy and Clinical Immunology.* 108:
    867–870.
Gendel, S.M. 2012. Comparison of international food allergen labeling regulations. *Regulatory
    Toxicology and Pharmacology.* 63: 279–285.

————. 2014. Food allergen risk management in the United States and Canada. In *Risk manage-ment for food allergy*, ed. C.B. Madsen, R.W.R. Crevel, C. Mills, and S.L. Taylor, 145–165. Waltham, NJ: Academic Press.

Gendel, S.M., and J. Zhu. 2013. Analysis of U.S. Food and Drug Administration food allergen recalls after implementation of the food allergen labeling and consumer protection act. *Journal of Food Protection.* 76: 1933–1938.

Gendel, S.M., N. Khan, and M. Yajnik. 2013. A survey of food allergen control practices in the U.S. food industry. *Journal of Food Protection.* 76: 302–306.

GMA (Grocery Manufacturers Association). 2009. Managing allergens in food processing establishments. Available at: http://americanbakers.org/wp-content/uploads/2012/10/GMA-ManagingAllergens9_09.pdf. Accessed 20 March 2017.

Gupta, R., D. Holdford, L. Bilaver, A. Dyer, J.L. Holl, and D. Meltzer. 2013. The economic impact of childhood food allergy in the United States. *JAMA Pediatrics.* 167: 1026–1031.

Hale, K. R. 2017. Undeclared allergens in FSIS-regulated products: Analysis of voluntary product recalls. Food Safety and Inspection Service allergen public meeting, March 16, 2017. Available at: https://www.fsis.usda.gov/wps/wcm/connect/e1eefd31-08a8-4e33-940f-a81fbf74c84a/Allergens-Slides-RobertsonHale-Seys-031617.pdf?MOD=AJPERES. Accessed 20 March 2017.

Hattersley, S., and R. King. 2014. How to keep allergic consumers happy and safe. In *Risk manage-ment for food allergy*, ed. C.B. Madsen, R.W.R. Crevel, C. Mills, and S.L. Taylor, 189–200. Waltham, NJ: Academic Press.

Hefle, S.L., J. Nordlee, and S.L. Taylor. 1996. Allergenic foods. *Critical Reviews in Food Science and Nutrition.* 36: 69–89.

Hefle, S.L., T.J. Furlong, L. Niemann, H. Lemon-Mule, S. Sicherer, and S.L. Taylor. 2007. Consumer attitudes and risks associated with packaged foods having advisory labeling regard-ing the presence of peanuts. *Journal of Allergy and Clinical Immunology.* 120: 171–176.

ICN (Institute of Child Nutrition). 2014. Managing food allergies in school nutrition programs. Available at: http://www.theicn.org/ResourceOverview.aspx?ID=507. Accessed 20 April 2017.

Jackson, L., F. Al-Taher, M. Moorman, J. Devries, T. Roger, K. Swanson, T. Fu, R. Salter, G. Dunaif, S. Estes, S. Albillos, and S. Gendel. 2008. Cleaning and other control and valida-tion strategies to prevent allergen cross-contact in food-processing operations. *Journal of Food Protection.* 71: 445–458.

Jackson, K. D., L. D. Howie, and L. J. Akinbami. 2013. Trends in allergic conditions among chil-dren: United States, 1997-2011. National Center for Health Statistics data brief. Available at: https://www.cdc.gov/nchs/data/databriefs/db121.pdf. Accessed 20 March 2017

Khuda, S.E., G.M. Sharma, D. Gaines, A.B. Do, M. Pereira, M. Chang, M. Ferguson, and K.M. Williams. 2016. Survey of undeclared egg allergen levels in the most frequently recalled food types (including products bearing precautionary labelling). *Food Additives and Contaminants: Part A.* 33: 1265–1273.

Lee, M.Y., and H. Xu. 2015. Food allergy knowledge, attitudes, and preparedness among restau-rant managerial staff. *Journal of Foodservice Business Research.* 18: 454–468.

Leitch, I.S., and J. McIntosh. 2014. The importance of food allergy training for environmen-tal health service professionals. In *Risk management for food allergy*, ed. C.B. Madsen, R.W.R. Crevel, C. Mills, and S.L. Taylor, 207–213. Waltham, NJ: Academic Press.

Marchisotto, M.J., L. Harada, O. Kamdar, B.M. Smith, S. Waserman, S. Sicherer, K. Allen, A. Muraro, S. Taylor, and R.S. Gupta. 2017. Food allergen labeling and purchasing habits in the United States and Canada. *Journal of Allergy and Clinical Immunology: In Practice.* 5: 345–351.

McEnrue, M., and V. Procopio. 2016. Survey of state epinephrine entity stocking laws. Available at: https://www.networkforphl.org/_asset/tmdxgd/50-State-Survey-Epinephrine-Entity-Stocking-Laws.pdf. Accessed 20 March 2017.

Muñoz-Furlong, A., and C.C. Weiss. 2009. Characteristics of food-allergic patients placing them at risk for a fatal anaphylactic episode. *Current Allergy and Asthma Reports.* 9: 57–63.

NASEM (National Academy of Sciences, Engineering and Medicine). 2016. Finding a path to safety in food allergy: Assessment of the global burden, causes, prevention, management, and public policy. Available at: http://www.nationalacademies.org/hmd/Reports/2016/finding-a-path-to-safety-in-food-allergy.aspx. Accessed 20 March 2017.

Nordlee, J.A., S.L. Taylor, J.A. Townsend, L.A. Thomas, and R.K. Bush. 1996. Identification of a Brazil-nut allergen in transgenic soybeans. *The New England Journal of Medicine*. 334: 688–692.

Nowak-Wegrzyn, A., M.K. Conover-Walker, and R.A. Wood. 2001. Food-allergic reactions in schools and preschools. *Archives of Pediatrics and Adolescent Medicine*. 155: 790–795.

NRA (National Restaurant Association). 2014. Restaurant industry forecast 2014. Available at: http://www.restaurant.org/Downloads/PDFs/News-Research/research/RestaurantIndustryForecast2014.pdf. Accessed 20 March 2017.

———. 2017. ServSafe allergen training. Available at: https://www.servsafe.com/allergens/the-course. Accessed 20 March 2017.

Pieretti, M.M., D. Chung, R. Pacenza, T. Slotkin, and S.H. Sicherer. 2009. Audit of manufactured products: Use of allergen advisory labels and identification of labeling ambiguities. *Journal of Allergy and Clinical Immunology*. 124: 337–341.

Radke, T.J., L.G. Brown, B. Faw, N. Hedeen, B. Matis, P. Perez, B. Viveiros, and D. Ripley. 2017. Restaurant food allergy practices - Six selected sites, United States, 2014. *Morbidity and Mortality Weekly Report*. 66: 404–407.

Sampson, H.A. 2004. Update on food allergy. *Journal of Allergy and Clinical Immunology*. 113: 805–819.

Sampson, M.A., A. Muñoz-Furlong, and S.H. Sicherer. 2006. Risk-taking and coping strategies of adolescents and young adults with food allergy. *Journal of Allergy and Clinical Immunology*. 117: 1440–1445.

Schellpfeffer, N.R., H.L. Leo, M. Ambrose, and A.N. Hashikawa. 2017. Food allergy trends and epinephrine autoinjector presence in summer camps. *Journal of Allergy and Clinical Immunology: In Practice*. 5: 358–362.

Sicherer, S.H., T.J. Furlong, J. DeSimone, and H.A. Sampson. 2001a. The US peanut and tree nut allergy registry: Characteristics of reactions in schools and day care. *Journal of Pediatrics*. 138: 560–565.

Sicherer, S.H., T.J. Furlong, A. Muñoz-Furlong, A.W. Burks, and H.A. Sampson. 2001b. A voluntary registry for peanut and tree nut allergy: Characteristics of the first 5149 registrants. *Journal of Allergy and Clinical Immunology*. 108: 128–132.

Stone, W.E., and J.M. Yeung. 2010. Principles and practices for allergen management and control in processing. In *Allergen management in the food industry*, ed. J.I. Boye and S.B. Godefroy, 145–165. Hoboken, NJ: John Wiley & Sons, Inc.

Taylor, S.L., and S.L. Hefle. 2005. Allergen control. *Food Technology*. 59 (40-43): 75.

———. 2006. Food allergen labeling in the USA and Europe. *Current Opinion in Allergy and Clinical Immunology*. 6: 186–190.

Turner, P.J., M.H. Gowland, V. Sharma, D. Ierodiakonou, N. Harper, T. Garcez, R. Pumphrey, and R.J. Boyle. 2014. Increase in anaphylaxis-related hospitalizations but no increase in fatalities: An analysis of United Kingdom national anaphylaxis data, 1992-2012. *Journal of Allergy and Clinical Immunology*. 135: 956–963.

US Code. 2004. Food Allergen Labeling and Consumer Protection Act of 2004 (Title II of Public Law 108-282, Title II). Available at: https://www.fda.gov/downloads/Food/GuidanceRegulation/UCM179394.pdf. Accessed 20 March 2017.

Weiss, C.C., and A. Muñoz-Furlong. 2008. Fatal food allergy reactions in restaurants and foodservice establishments: Strategies for prevention. *Food Protection Trends*. 28: 657–661.

Wen, H., and J. Kwon. 2016. Food allergy risk communication in restaurants. *Food Protection Trends*. 36: 372–383.

White, M.V., S.L. Hogue, D. Odom, D. Cooney, J. Bartsch, D. Goss, K. Hollis, C. Herrem, and S. Silvia. 2016. Anaphylaxis in schools: Results of the EPIPEN4SCHOOLS survey combined analysis. *Pediatric Allergy, Immunology, and Pulmonology*. 29: 149–154.

Chapter 2
A Review of the Distribution and Costs of Food Allergy

Ruchi S. Gupta, Alexander M. Mitts, Madeline M. Walkner, and Alana Otto

2.1 Introduction

Food allergy is a significant disease and requires attention in both medicine and society. It affects 8% of U.S. children (Gupta et al. 2011) and there is no established cure for food allergy yet. Moreover, potential allergen exposure and the risk of severe allergic reactions are part of daily life. Epinephrine auto-injectors (EAIs) are the only approved treatment for severe allergic reactions, and access to these devices among food-allergic children is not always possible. Food-allergic people and their caretakers are forced to be constantly vigilant, often at great psychological and financial cost.

Food allergy is more likely to occur in African American and Asian children than White children, and the rate of reactions requiring emergency department (ED) visits or hospitalizations is growing fastest among Hispanic children (Dyer et al. 2015). Importantly, formal diagnoses and access to treatment are less likely among racial/ethnic minorities than among White children. Lower income families also have limited access to preventative measures but spend more than twice what higher-income families do on emergency department visits for food-allergic reactions. Urban children

R.S. Gupta (✉) • A.M. Mitts
Northwestern University Feinberg School of Medicine,
750 N. Lake Shore Drive 6th FL, Chicago, IL 60611, USA
e-mail: r-gupta@northwestern.edu

M.M. Walkner
Ann & Robert H. Lurie Children's Hospital of Chicago, Chicago, IL 60611, USA

A. Otto
Northwestern University Feinberg School of Medicine,
750 N. Lake Shore Drive 6th FL, Chicago, IL 60611, USA

Ann & Robert H. Lurie Children's Hospital of Chicago, Chicago, IL 60611, USA

© Springer International Publishing AG 2018
T.-J. Fu et al. (eds.), *Food Allergens*, Food Microbiology and Food Safety,
DOI 10.1007/978-3-319-66586-3_2

have also been shown to have higher numbers of food allergy diagnosis and higher rates of ED visits than those living the in suburbs. These disparities, together with the large size of the affected population, make childhood food allergy a serious public health problem. Increasing public access to information about food allergy and its treatment is vital.

It has been demonstrated that the quality of life among parents of children with food allergy is worse than that of parents of non-allergic children. This difference exists even among parents who feel comfortable with controlling their children's food allergies. Avenues to improve quality of life among food-allergic children and their caregivers have been largely unexplored. At the consumer level, there is a great deal of ambiguity in the regulation and meaning of allergen warnings on packaged foods, which may lead to misinformed and potentially dangerous food purchases. From an economic standpoint, food allergy costs the US \$24.8 billion annually (Gupta et al. 2013); much of this cost is borne by the families of food-allergic children (Gupta et al. 2013).

Schools exist at the intersection of the many spheres discussed above and represent an opportunity to improve outcomes for food-allergic children around the country. Chicago Public Schools (CPS) is the largest school system to date to implement a program to make undesignated epinephrine available for any student experiencing anaphylaxis. This program has potentially saved dozens of lives but has also cast into sharp relief about the disparities in and necessity of access to EAIs. The data gleaned from the CPS initiative should motivate us to increase safety and awareness around food allergy.

This chapter will aim to make clear the state of food allergy in the U.S., and on which parties the burden of the disease lies. The studies discussed herein aimed to gather and organize population-level information on the prevalence, distribution, and cost of food allergy, as well as to understand some of the social and psychological responses to the challenges of food allergy.

2.2   The Prevalence, Severity, and Distribution of Childhood Food Allergy in the U.S.

There is a lack of comprehensive data on both the number of children in the U.S. living with food allergy, and their demographics. Efforts to characterize the scope of the disease have been limited by small and non-representative samples as well as by the use of non-standardized diagnostic criteria (Gupta et al. 2011). We therefore conducted a population-based, cross-sectional survey of a large representative population of U.S. children ($n = 38,480$) in an attempt to define the prevalence and severity of pediatric food allergy in the U.S. Recruitment employed a dual-sample approach, in which a probability-based sample statistically representative of U.S. households with children was used to identify and correct for a sampling and non-sampling bias introduced by a larger, opted-in online sample (Gupta et al. 2011). In this survey, prevalence estimates include report of both *convincing* allergy, defined as participant report of a food allergy plus a history of one or more common symptoms, and *confirmed* allergy, defined by the criteria of a convincing allergy plus confirmatory physician diagnosis by serum specific immunoglobulin E (IgE) testing, skin prick testing (SPT), or oral food challenge (OFC).

2.2.1   Prevalence

Based on our primary research survey (Table 2.1), 8% of U.S. children were found to have food allergy. Thirty-four percent of these children, or approximately 2.5% of the total population, were allergic to more than one food and were therefore at an increased risk of severe reaction. Additionally, 3.1% of all children and 39% of food-allergic children were found to have *severe* food allergy, defined as a history of at least one reaction of one or more of the following symptoms: anaphylaxis, hypotension, trouble breathing, or wheezing. Males were significantly more likely than females to have severe allergies. There were no significant differences between genders in the frequencies of convincing allergy or confirmed allergy.

2.2.1.1   Prevalence by Age

The overall prevalence of food allergy varied significantly by age and was highest among children 3–5 years old (9.2%) (Table 2.1). Significant variance in prevalence according to age was observed for peanut, shellfish, tree nut, wheat, and egg allergy. Adolescents were at significantly higher risk for severe allergic reactions than were children aged 0–2 (odds ratio [OR] = 2.1). The odds of having a confirmed food allergy did not vary significantly with age (Gupta et al. 2011).

2.2.1.2   Prevalence by Allergen

Peanut was the most common food allergen, with 2% of all children and of 25% of food-allergic children in this survey allergic to peanuts. This estimate is two times higher than that made by a Canadian study (Ben-Shoshan et al. 2010). The prevalence of fin fish allergy (0.5%), was also higher than the previously reported 0.3% of children and adults (Ben-Shoshan et al. 2010). The prevalence of other common food allergens assessed in this survey was consistent with previous findings (Gupta et al. 2011). Severe reactions were most common among children with tree nut, peanut, shellfish, soy, and fin fish allergies (Gupta et al. 2011).

2.2.1.3   Prevalence by Race/Ethnicity

African American and Asian children were significantly more likely than White children to have food allergies (OR = 1.8 and 1.4, respectively) but less likely to have physician-confirmed diagnoses (OR = 0.8 and 0.7, respectively). Hispanic children were also significantly less likely than White children to have physician-confirmed allergies (OR = 0.8) (Gupta et al. 2011).

Table 2.1 Prevalence of common food allergies according to age group

Age group	Frequency, % (95% CI)									
	All allergens (N = 3339)	Peanut (N = 767)	Milk (N = 702)	Shellfish (N = 509)	Tree Nut (N = 430)	Egg (N = 304)	Fin Fish (N = 188)	Strawberry (N = 189)	Wheat (N = 170)	Soy (N = 162)
Prevalence among all children surveyed										
All ages (N = 38 480)	8.0 (7.7–8.3)	2.0 (1.8–2.2)	1.7 (1.5–1.8)	1.4 (1.2–1.5)	1.0 (0.9–1.2)	0.8 (0.7–0.9)	0.5 (0.4–0.6)	0.4 (0.4–0.5)	0.4 (0.3–0.5)	0.4 (0.3–0.4)
0–2 years (n = 5429)	6.3 (5.6–7.0)	1.4 (1.1–1.8)	2.0 (1.6–2.4)	0.5 (0.3–0.8)	0.2 (0.2–0.5)	1.0 (0.7–1.3)	0.3 (0.1–0.4)	0.5 (0.3–0.7)	0.3 (0.1–0.5)	0.3 (0.2–0.4)
3–5 years (n = 5910)	9.2 (8.3–10.1)	2.8 (2.3–3.4)	2.0 (1.7–2.5)	1.2 (0.8–1.6)	1.3 (1.0–1.7)	1.3 (0.9–1.7)	0.5 (0.3–0.8)	0.5 (0.3–0.8)	0.5 (0.3–0.7)	0.5 (0.3–0.7)
6–10 years (n = 9911)	7.6 (7.0–8.2)	1.9 (1.6–2.3)	1.5 (1.2–1.8)	1.3 (1.1–1.6)	1.1 (0.87–1.4)	0.8 (0.6–1.1)	0.5 (0.3–0.7)	0.4 (0.3–0.5)	0.4 (0.3–0.5)	0.3 (0.2–0.5)
11–13 years (n = 6716)	8.2 (7.4–9.0)	2.3 (1.9–2.8)	1.4 (1.1–1.8)	1.7 (1.3–2.1)	1.2 (1.0–1.6)	0.5 (0.4–0.8)	0.6 (0.4–0.8)	0.4 (0.3–0.6)	0.7 (0.5–0.9)	0.6 (0.4–0.8)
≥14 years (n = 10 514)	8.6 (7.9–9.3)	1.7 (1.4–2.1)	1.6 (1.3–1.9)	2.0 (1.7–2.5)	1.2 (0.9–1.5)	0.4 (0.2–0.5)	0.6 (0.4–0.9)	0.4 (0.3–0.6)	0.3 (0.2–0.4)	0.3 (0.2–0.4)
P	0.0000	0.0001	0.0504	0.0000	0.0000	0.0000	0.1045	0.7700	0.0089	0.0509
Prevalence among children surveyed with food allergy										
All ages (N = 3339)	—	25.2 (23.3–27.1)	21.1 (19.4–22.8)	17.2 (15.6–18.9)	13.1 (11.7–14.6)	9.8 (8.5–11.1)	6.2 (5.2–7.3)	5.3 (4.4–6.3)	5.0 (4.2–6.0)	4.6 (3.8–5.6)
0–2 years (n = 469)	—	22.2 (17.4–27.8)	31.5 (26.6–36.8)	7.5 (4.7–11.9)	5.4 (3.6–8.1)	15.8 (12.0–20.4)	4.0 (2.3–6.9)	7.5 (5.2–8.2)	4.0 (2.2–7.2)	4.2 (2.7–6.5)
3–5 years (n = 539)	—	30.3 (25.8–35.3)	22.1 (18.3–26.5)	12.9 (9.7–16.9)	14.3 (11.1–18.2)	13.7 (10.5–17.6)	5.7 (3.8–8.6)	5.5 (3.6–8.2)	5.0 (3.2–7.7)	5.1 (3.3–7.8)

6–10 years (n = 847)	—	25.5 (22.0–29.5)	19.6 (16.6–23.0)	17.1 (14.0–20.6)	14.3 (11.6–17.5)	11.1 (8.6–14.3)	6.2 (4.5–8.5)	4.8 (3.4–6.9)	5.0 (3.5–7.1)	4.0 (2.6–6.2)
11–13 years (n = 584)	—	28.1 (23.7–32.9)	17.7 (14.2–22.0)	20.4 (16.8–24.7)	15.2 (12.0–19.2)	6.6 (4.4–9.9)	7.0 (4.8–10.1)	4.6 (3.1–6.8)	8.2 (5.9–11.2)	6.9 (4.7–10.0)
2:14 years (n = 900)	—	20.2 (17.0–23.7)	18.4 (15.3–22.1)	23.8 (20.1–27.9)	13.4 (10.7–16.6)	4.1 (2.9–5.9)	7.2 (5.2–9.8)	4.9 (3.3–7.3)	3.3 (2.1–5.0)	0.3 (0.2–0.4)
P	—	0.0050	0.0001	0.0000	0.0010	0.0000	0.4646	0.4486	0.0174	0.1296

Common food allergens are those reported with a frequency of $n > 150$

Gupta et al. (2011)

2.2.1.4   Prevalence by Socioeconomic Status

The odds of having food allergy were significantly lower for children in households with annual incomes of <$50,000 versus ≥$50,000 (OR = 0.5). Children in house-holds with annual incomes of <$50,000 were also significantly less likely to have confirmed diagnoses (OR = 0.5) and severe allergies (OR = 0.8) (Gupta et al. 2011).

2.2.1.5   Prevalence by Geography

The odds of having food allergy were significantly higher for children from the Northeast (OR = 1.3), South (OR = 1.5), and West (OR = 1.3) regions of the U.S. versus children from the Midwest. There was no significant difference between regions in terms of the odds of having confirmed (versus convincing) allergy or severe (versus non-severe) allergy (Gupta et al. 2012).

2.2.2   What the Prevalence, Severity, and Distribution of Childhood Food Allergy Tell Us

Eight percent of children in this study had food allergy; this equates to an estimated 5.9 million affected children in the U.S. This prevalence is higher than many previous estimates and underscores the importance of food allergy as a public health concern. Furthermore, in this study, 39% of food-allergic children had severe food allergies, and 34% had multiple food allergies. To our knowledge, the prevalence of severe food allergy among a representative sample of U.S. children has not been previously reported. Importantly, the distribution of food allergy varies significantly among racial and socioeconomic groups as well as by geographic region. While food allergy was found to be more common among African American and Asian children, these children were less likely than their White counterparts to have physician-confirmed diagnoses. This difference is likely influenced by differences in healthcare access and utilization between these populations. Further work is needed to determine how bio-logical, social, and economic factors influence the incidence as well as the diagnosis and management of food allergy in patients with different racial and socioeconomic backgrounds (Gupta et al. 2011).

2.3   Geographic Variability of Food Allergy in the U.S.

The same population-based, cross-sectional survey was used to determine the geographic distribution of food allergy prevalence in the U.S. (Gupta et al. 2012). Data were analyzed for 38,465 children. Geographic characteristics assessed included state, latitude, zip code, and urban/rural status. Latitude was assigned by zip code, and latitudes were collapsed into terciles: northern ($\geq$41.8° N latitude), middle (34.3° N to 41.7° N latitude), and southern ($\leq$34.2° N latitude). Urban/rural status was also determined by zip code and classified by the following designations, in order of decreasing population density: urban center, metropolitan city, urban outskirt, suburban area, small town, rural area. The primary outcome measure was food allergy prevalence. The prevalence of severe food allergy (defined in Sect. 2.2.1) was also measured. Multiple logistic regression models, adjusted for race/ethnicity, gender, age, household income, and latitude, were used to assess associations between geographic variables and the presence and severity of food allergy.

2.3.1   Food Allergy by Latitude

Odds of food allergy were significantly higher in southern and middle latitudes than they were in northern latitudes (OR = 1.5 and 1.3, respectively). The gradation of prevalence in our results suggests a north-to-south increase in the rate of food allergy. Odds of severe food allergy did not vary significantly with geographic region (Gupta et al. 2012).

2.3.2   Food Allergy by Urban/Rural Status

Prevalence of food allergy varied significantly with urban/rural status (Tables 2.2 and 2.3). Increasingly urban settings corresponded with increasing prevalence of food allergy, which ranged from 6.2% in rural areas to 9.8% in urban centers. Prevalence rates for many specific allergies varied significantly with population density. Only milk and soy allergies did not significantly vary with geographic area. Peanut was consistently among the two most common allergens and was the most common in all but rural areas, where it was replaced by milk allergy. The odds ratio for having food allergy was highest in urban versus rural environments. Odds of severe food allergy did not differ significantly by urban/rural status.

2.3.3   What the Geographic Distribution of Food Allergy in the U.S. Tells Us

To our knowledge, an urban/rural difference in food allergy prevalence in the U.S. has not been previously reported. This information contributes broad demographic information to the goal of understanding the etiology and impact of food allergy and may ultimately guide development of treatments. Importantly, urban/rural status may affect allergy prevalence but not morbidity (Gupta et al. 2012).

Table 2.2 Food allergy prevalence by geographic area: Overall and by common allergen

Area	Frequency, % (95% Confidence Interval)								
	All allergens	Peanut	Shellfish	Milk	Fin Fish	Egg	Tree Nut	Wheat	Soy
Urban centers	9.8	2.8	2.4	1.8	1.8	1.3	1.2	0.8	0.6
	(8.6–11.0)	(2.2–3.5)	(1.8–3.0)	(1.4–2.4)	(1.4–2.3)	(0.9–1.8)	(0.8–1.6)	(0.5–1.1)	(0.3–0.9)
Metro cities	9.2	2.4	1.4	1.8	0.9	1.0	1.3	0.9	0.4
	(8.4–10.1)	(2.0–2.9)	(1.1–1.8)	(1.5–2.2)	(0.6–1.2)	(0.7–1.3)	(1.0–1.7)	(0.7–1.2)	(0.3–0.6)
Urban outskirts	7.8	1.8	1.5	1.4	0.8	0.5	1.0	0.4	0.4
	(7.0–8.6)	(1.5–2.3)	(1.2–2.0)	(1.1–1.7)	(0.5–1.1)	(0.4–0.8)	(0.8–1.3)	(0.3–0.6)	(0.2–0.6)
Suburban areas	7.6	2.0	1.2	1.5	0.7	0.7	1.2	0.8	0.3
	(6.9–8.2)	(1.7–2.4)	(1.0–1.5)	(1.2–1.8)	(0.5–0.9)	(0.5–0.9)	(0.9–1.5)	(0.6–1.0)	(0.2–0.5)
Small towns	7.2	1.6	1.0	1.4	0.5	0.7	0.9	1.1	0.5
	(5.7–8.6)	(1.0–2.6)	(0.6–1.7)	(0.9–2.3)	(0.3–1.0)	(0.4–1.4)	(0.6–1.6)	(0.7–1.9)	(0.2–0.9)
Rural areas	6.2	1.3	0.8	1.5	0.2	0.5	0.6	0.5	0.2
	(5.6–6.8)	(1.0–1.6)	(0.6–1.1)	(1.2–1.8)	(0.1–0.4)	(0.3–0.7)	(0.4–0.8)	(0.3–0.7)	(0.1–0.4)
P	<0.0001	<0.0001	<0.0001	0.3993	<0.0001	0.0045	0.0001	0.0040	0.2658

Urban/rural status was assigned by zip code using the Rural-Urban Commuting Area Codes (RUCA) version 2.0
Gupta et al. (2012)

Table 2.3 Odds of food allergy and severe versus mild/moderate food allergy by geographic area, adjusted for race/ethnicity, gender, age, household income, and latitude

Area versus rural	Odds of food allergy		Odds of severe food allergy	
	Unadjusted	Adjusted	Unadjusted	Adjusted
Urban centers	1.7 (1.5–2.0)	1.5 (1.3–1.8)	1.4 (1.0–1.8)	1.3 (0.9–1.8)
Metro cities	1.5 (1.3–1.7)	1.4 (1.2–1.6)	1.1 (0.9–1.5)	1.1 (0.8–1.4)
Urban outskirts	1.3 (1.1–1.5)	1.2 (1.1–1.4)	1.0 (0.8–1.3)	1.0 (0.8–1.3)
Suburban areas	1.2 (1.1–1.4)	1.2 (1.0–1.3)	1.1 (0.9–1.4)	1.0 (0.8–1.3)
Small towns	1.2 (0.9–1.4)	1.2 (0.9–1.5)	1.2 (0.8–1.8)	1.1 (0.7–1.7)

Gupta et al. (2012)

2.4   Food Allergy Sensitization and Presentation in Siblings of Children with Food Allergic

Parents of food-allergic children are often concerned about the risk of food allergy in their other children, and may ask about screening their asymptomatic children for food allergies. Little is known about the prevalence of sensitization and true food allergy in the siblings of food-allergic children or about the utility of screening asymptomatic siblings for sensitization or allergies to common food allergens. We therefore aimed to determine the prevalence of both sensitization and true food allergy among siblings of food-allergic children by evaluating a cohort of children with confirmed food allergy ($n = 478$) and their siblings ($n = 642$) (Gupta et al. 2015). Siblings were evaluated for laboratory evidence of food sensitization to nine common allergens using total and specific serum IgE as well as skin prick testing. Sensitization was defined as positive IgE and/or SPT in the absence of clinical symptoms of allergy. True food allergy was defined as positive IgE and/or SPT plus clinical symptoms consistent with food allergy.

2.4.1   Prevalence of Sensitization and True Allergy in Siblings of Food-Allergic Children

Approximately one third (33.4%) of siblings of children with food allergy had neither sensitization nor clinical symptoms to the foods tested. Approximately one half (53%) were sensitized to one or more foods, most commonly wheat (37%), milk (35%), and egg (35%). Thirteen percent of siblings were diagnosed with true food allergy. The most common true allergens among siblings of food-allergic children were milk (5.9%), egg (4.4%), and peanut (3.7%).

2.4.2 What the Prevalence and Sensitization of Food Allergy in Siblings of Food-Allergic Children Tells Us

In this study, the prevalence of true food allergy among siblings of food-allergic children was low. We therefore recommend against withholding foods containing common allergens from these patients in the absence of symptoms suggestive of allergy. We also recommend against the routine screening of asymptomatic siblings of food-allergic children, as the presence of laboratory sensitization alone is insufficient to diagnose food allergy, and misdiagnosis may lead to unnecessary elimination of foods from the sibling's diet (Gupta et al. 2015).

2.5 Pediatric Emergency Department Visits and Hospitalizations for Food-Induced Anaphylaxis in Illinois

Little is known about the frequency with which food-allergic children access emergency departments or are admitted to hospitals for acute allergic reactions; similarly, there is a paucity of data about the effects of race, ethnicity, and socioeconomic status on food allergy-related healthcare utilization. We therefore sought to quantify food allergy-related ED visits and hospitalizations by children from various racial and socioeconomic backgrounds using medical record data from 2008 to 2012 in Illinois (Dyer et al. 2015).

2.5.1 Pediatric ED Visits and Hospitalizations Over Time

The average annual rate of ED visits and hospital admissions for food-induced anaphylaxis among Illinois children over the 5-year study period was 10.9 per 100,000 (See Table 2.4 for demographic breakdown). Eleven percent of children who presented to an ED for food-induced anaphylaxis during the study period were hospitalized. Between 2008 and 2012, the rate of ED visits for food-induced anaphylaxis increased from 6.3 to 17.2 per 100,000 ($p < 0.001$), with an annual increase of 29%. The rate of hospitalizations for food-induced anaphylaxis increased 19% per year over this period, from 0.8 per 100,000 in 2008 to 1.5 per 100,000 in 2012 ($p < 0.001$). Increases in rates of ED visits and hospitalizations were seen among children of all ages, sexes, races/ethnicities, insurance types, and metropolitan statuses. The largest annual percent increases were seen among Hispanic children (44%, $p < 0.01$), children with public insurance (30%, $p < 0.01$), and children from urban neighborhoods outside Chicago (49%, $p < 0.01$). (Dyer et al. 2015).

Table 2.4 Rates of ED visits and hospitalization for food-induced anaphylaxis in Illinois 2008–2012

Variable	Rate of ED visits and hospital admissions for food-induced anaphylaxis per 100,000 children (95% CI)				
	2008 (n = 226)	2009 (n = 279)	2010 (n = 319)	2011 (n = 481)	2012 (n = 590)
Overall	6.3 (5.5–7.2)	7.8 (6.9–8.8)	9.1 (8.2–10.2)	13.9 (12.7–15.2)	17.2 (15.9–18.7)
Age group, year					
0–4 (n = 840)	11.9 (9.7–14.4)	15.0 (12.6–17.8)	16.8 (14.1–19.8)	25.5 (22.1–29.2)	30.5 (26.9–34.5)
5–9 (n = 419)	4.6 (3.3–6.2)	6.5 (4.9–8.4)	7.2 (5.5–9.2)	12.8 (10.5–15.4)	17.7 (15.0–20.7)
10–14 (n = 284)	3.4 (2.3–4.9)	5.1 (3.7–6.9)	5.7 (4.2–7.5)	8.5 (6.7–10.7)	9.9 (8.0–12.2)
15–19 (n = 351)	5.3 (4.0–7.1)	4.8 (3.5–6.4)	7.3 (5.6–9.2)	9.4 (7.5–11.6)	11.8 (9.7–14.2)
Sex					
Male (n = 1117)	7.2 (6.1–8.6)	9.9 (8.5–11.5)	11.4 (9.9–13.1)	15.2 (13.5–17.2)	18.9 (17.0–21.1)
Female (n = 777)	5.4 (4.4–6.6)	5.6 (4.6–6.9)	6.7 (5.6–8.1)	12.5 (10.8–14.3)	15.5 (13.7–17.5)
Race/ethnicity					
Asian, non-Hispanic (n − 124)	12.9 (7.8 20.2)	11.6 (6.7 18.5)	15.2 (9.8 22.7)	22.4 (15.3 31.6)	24.1 (16.9 33.3)
Black, non-Hispanic (n = 369)	8.0 (5.9–10.6)	9.4 (7.1–12.2)	9.2 (6.9–12.0)	17.5 (14.2–21.3)	20.2 (16.8–24.4)
White, non-Hispanic (n = 1009)	6.0 (5.0–7.3)	7.6 (6.4–8.9)	10.1 (8.7–11.7)	14.2 (12.5–16.0)	16.8 (15.0–18.8)
Hispanic (n = 248)	2.8 (1.8–4.2)	3.8 (2.6–5.3)	4.5 (3.1–6.2)	7.4 (5.6–9.5)	12.5 (10.2–15.2)
Insurance type					
Private insurance (n = 1374)	7.7 (6.6–8.9)	8.3 (7.2–9.6)	11.0 (9.7–11.2)	16.9 (15.3–18.8)	18.8 (17.1–20.8)
Public insurance (n = 519)	3.9 (2.9–5.1)	7.0 (5.6–8.6)	6.0 (4.7–7.5)	9.2 (7.7–10.9)	14.8 (12.8–17.0)
Metropolitan status					
Chicago, urban (n = 639)	11.0 (8.7–13.8)	14.0 (11.4–17.1)	14.0 (11.4–17.8)	24.7 (21.2–28.7)	27.6 (23.9–31.7)
Chicago, not urban (n = 978)	7.4 (6.2–8.9)	8.5 (7.2–10.0)	10.4 (9.0–12.1)	13.5 (11.8–15.4)	17.8 (15.9–19.9)
Outside Chicago, urban (n = 125)	3.2 (1.8–5.9)	2.7 (1.3–4.9)	5.3 (3.2–8.2)	9.8 (6.9–13.5)	12.2 (9.1–16.2)
Outside Chicago, not urban (n = 142)	1.4 (0.7–2.5)	3.7 (2.5–5.4)	3.3 (2.1–4.9)	5.2 (3.7–7.2)	5.9 (4.3–7.9)

(continued)

Table 2.4 (continued)

Variable	Rate of ED visits and hospital admissions for food-induced anaphylaxis per 100,000 children (95% CI)				
	2008 (*n* = 226)	2009 (*n* = 279)	2010 (*n* = 319)	2011 (*n* = 481)	2012 (*n* = 590)
Hospitalization status					
Discharged from ED (*n* = 1753)	6.0 (5.3–6.9)	7.1 (6.2–8.0)	8.5 (7.5–9.5)	12.8 (11.6–14.0)	16.0 (14.7–17.4)
Admitted to hospital (*n* = 203)	0.8 (0.5–1.0)	1.0 (0.7–1.3)	1.2 (0.9–1.6)	1.4 (1.1–1.9)	1.5 (1.1–1.9)
Food allergen					
Peanut (*n* = 649)	2.2 (1.8–2.8)	2.2 (1.8–2.8)	3.7 (3.1–4.4)	4.8 (4.1–5.6)	5.6 (4.9–6.5)
Tree nut (*n* = 318)	0.9 (0.6–1.3)	1.5 (1.2–2.0)	1.5 (1.1–1.9)	2.3 (1.9–2.9)	2.9 (2.4–3.5)
Fin fish (*n* = 123)	0.4 (0.3–0.8)	0.7 (0.5–1.0)	0.4 (0.2–0.7)	0.9 (0.6–1.3)	1.1 (0.7–1.4)
Milk (*n* = 103)	0.4 (0.2–0.7)	0.4 (0.3–0.7)	0.3 (0.1–0.5)	0.8 (1.5–1.2)	1.0 (0.7–1.4)
Other food (*n* = 452)	1.8 (1.4–2.3)	2.0 (1.6–2.5)	2.0 (1.5–2.9)	3.1 (2.5–3.7)	4.1 (2.5–4.9)
Unknown food (*n* = 259)	0.6 (0.4–0.9)	1.0 (0.7–1.3)	1.3 (0.9–1.7)	2.0 (1.5–2.5)	2.6 (2.1–3.2)
Hospital type					
Dedicated pediatric hospital (*n* = 771)	2.6 (2.1–3.2)	3.5 (3.0–4.2)	3.4 (2.9–4.1)	5.7 (5.0–6.6)	6.8 (6.0–7.7)
Combined adult and pediatric hospital with PICU (*n* = 349)	1.5 (1.3–1.9)	1.2 (0.9–1.6)	1.9 (1.5–2.5)	2.3 (1.8–2.8)	3.1 (2.5–3.7)
Combined adult and pediatric hospital without PICU (*n* = 773)	2.2 (1.8–2.8)	3.1 (2.5–3.7)	3.7 (3.1–4.4)	5.8 (5.1–6.7)	7.4 (6.5–8.3)

Dyer et al. (2015)

2.5.2 Pediatric ED Visits and Hospitalizations by Patient Demographics

The highest rates of ED visits and hospitalizations were seen among children 0–4 years of age (12–30.5 per 100,000); however, the largest percent annual increase in visits (40%) was seen among children 5–9 years. Significantly more infants presenting to an ED for food-induced anaphylaxis were admitted (42%) than children 1 year or older (18%, $p = 0.02$) (Table 2.4). Hospital length of stay did not vary significantly by age (Dyer et al. 2015).

Differences in ED visit and hospitalization rates were also seen between races and ethnicities. The highest rates of ED visits and hospitalizations were seen among Asian children, while the lowest rates were seen among Hispanic children; however, as mentioned above, Hispanic children experienced the largest percent annual increase in visit rates (44%), while the percent annual increase was lowest among Asian children (21%) (Table 2.4). Rates of ED visits and hospitalizations as well as annual percent increases were similar between White and African American children. Hospital length of stay did not vary significantly by race or ethnicity (Dyer et al. 2015).

Variation in ED visit and hospitalization rates by socioeconomic and metropolitan statuses was also seen. ED visits and hospitalizations were significantly more frequent among those with private insurance (7.7–18.8 per 100,000) than among those with public insurance (3.9–14.8 per 100,000), and the annual percent increase was higher among privately insured children (39% versus 30%). Rates of ED visits and hospitalization were highest among children in urban Chicago neighborhoods, while children in suburban neighborhoods outside Chicago visited least frequently (Table 2.4). Annual percent increases in rates for all visit types were significantly increased for all metropolitan statuses, and children from urban neighborhoods outside Chicago had the most pronounced annual percent increase (49%). The highest rate of hospitalization was seen among children in suburban Chicago neighborhoods. Hospital length of stay did not vary significantly by insurance type or metropolitan status (Dyer et al. 2015).

Rates of ED visits and hospitalizations also varied by specific allergen. The most frequent overall rates of ED visits and hospitalizations were seen among children with peanut allergy (Table 2.4), while hospitalization was most frequent among children with milk-induced anaphylaxis (Dyer et al. 2015). The annual percent increase in ED visits and hospitalizations was most pronounced for children with tree-nut induced anaphylaxis. Hospital length of stay did not vary significantly by allergen (Dyer et al. 2015).

2.5.3   What Data on Pediatric ED and Hospital Visits Due to Food-Induced Anaphylaxis Tells Us

Food allergy-related ED visits and hospitalizations are increasing in frequency across socioeconomically and racially/ethnically diverse populations. Understanding these trends may help better target efforts aimed at preventing food-allergic reactions. Our work suggests the epidemiology of food allergy may be changing, as children with the lowest rates of ED visits and hospitalizations experienced the highest annual percent increases in visit rates over the study period. It is unclear whether these observations represent changes in disease prevalence or in healthcare access and utilization. In this study, children from urban Chicago neighborhoods visited most frequently, and those from suburban neighborhoods outside the city had the highest annual percent increase in visits. The first of these findings, at least, is consistent with those presented in Sect. 2.2, where we found that urban status was positively correlated with the prevalence of food allergy. This study also found that children with peanut and tree nut allergies are more likely to visit the ED than children with other food allergies, which is in keeping with previous literature reporting that children with peanut and tree nut allergies are more likely to have severe reactions (Dyer et al. 2015).

2.6   Differences in Empowerment and Quality of Life Among Parents of Children with Food Allergy

Given that food allergens are often difficult to avoid, and treatment for food allergy is limited, food allergies can strain relationships and demonstrably lower quality of life for allergic children and their families. We therefore aimed to describe the differences in empowerment to care for children with food allergies and food allergy-related quality of life (FAQOL) in mothers and fathers of a cohort of 876 children with food allergy (Warren et al. 2015). Empowerment was assessed through 16 items adapted from the Family Empowerment Scale (Koren et al. 1992). Empowerment scores derived largely from measures of parental confidence, parental involvement, perceived ability to act, and parental food allergy education. FAQOL was assessed through 15 items adapted from the Food Allergy-related Quality Of Life–Parental Burden scale (Cohen et al. 2004). The questionnaires (Table 2.5) used to measure both empowerment and FAQOL were negatively worded, meaning that a lower score represented greater empowerment or FAQOL. We also sought to understand how the relationships between perceived levels of support and resource access tracked with FAQOL. These relationships are summarized in Table 2.6.

2.6.1   Parental Empowerment and FAQOL

Significant differences were seen between mothers and fathers in both empowerment and FAQOL. Mothers reported significantly greater empowerment ($p < 0.001$) and significantly lower FAQOL ($p < 0.001$) than fathers, regardless of their children's allergy severity, specific allergen, or comorbidities. Empowerment was not significantly associated with FAQOL among mothers or fathers. Having "the support of friends and family to help care for your child" and "the resources you need to care for your child" were significant predictors of high FAQOL in both mother and fathers but had a significantly greater impact on mothers' FAQOL (Warren et al. 2015). Non-Hispanic White ethnicity was significantly associated with increased FAQOL among both mothers and fathers, while Hispanic ethnicity was associated with reduced FAQOL among fathers only. Child age was not associated with paternal FAQOL; however, high maternal FAQOL was positively associated with having a child 2–5, 6–10, and 11–13 years old. The presence of comorbid conditions was

Table 2.5 Parental food allergy-related empowerment and quality of life by child's food allergy type

	Children with peanut allergy			Children with milk allergy			Children with egg allergy			Children with tree nut allergy		
	Father	Mother	P value (father vs. mother)	Father	Mother	P value (father vs. mother)	Father	Mother	P value (fathers vs. mother)	Father	Mother	P value (father vs. mother)
Items for adapted Family Empowerment Scale	1.96	1.73	0.0168	2.01	1.85	0.0375	2.05	1.75	0.0251	2.04	1.85	0.2582
When problems arise with my child, I handle them pretty well	2.41	2.18	0.0045	2.51	2.23	0.0089	2.44	2.23	0.0120	2.40	2.14	0.0786
I know what to do when problems arise with my child	2.08	1.96	0.1714	2.12	2.09	0.5838	2.23	2.07	0.2750	2.17	1.94	0.0807
I feel like my family life is under control	2.03	1.61	0.0000	2.17	1.64	0.0000	2.19	1.60	0.0000	2.04	1.62	0.0015
I am able to make decisions about what my child needs medically	2.81	1.59	0.0000	2.94	1.61	0.0000	2.75	1.52	0.0000	2.55	1.61	0.0000
I make sure I stay in regular contact with physicians who are caring for my child	2.10	1.72	0.0000	2.18	1.81	0.0006	2.24	1.64	0.0000	2.09	1.71	0.0017
I believe I can solve problems with my child when they happen	1.87	1.64	0.0318	1.86	1.66	0.0295	1.94	1.63	0.0043	1.91	1.60	0.0078

(continued)

Table 2.5 (continued)

	Children with peanut allergy			Children with milk allergy			Children with egg allergy			Children with tree nut allergy		
	Father	Mother	P value (father vs. mother)	Father	Mother	P value (father vs. mother)	Father	Mother	P value (fathers vs. mother)	Father	Mother	P value (father vs. mother)
When dealing with my child, I focus on good things as well as the problems	1.75	1.65	0.5265	1.76	1.60	0.1218	1.79	1.63	0.3266	1.81	1.62	0.1084
I feel I am a good parent I can calmly handle a crisis situation involving my child	1.93	1.92	0.0926	1.93	2.02	0.0366	2.00	1.92	0.7533	2.09	1.99	0.9311
I am confident in my abilities to protect my child from danger	1.69	1.65	0.5983	1.74	1.72	0.8809	1.75	1.70	0.9377	1.72	1.74	0.9928
I trust my physician	1.91	1.65	0.0011	2.06	1.74	0.0000	1.95	1.70	0.0074	1.95	1.74	0.0988
I am decisive and act quickly	2.03	1.97	0.4728	2.11	2.08	0.4816	2.11	1.95	0.3541	2.13	2.04	0.8025
I feel confident in my ability to deal with my child's medical problems	2.02	1.67	0.0002	2.08	1.76	0.0001	2.17	1.74	0.0008	2.08	1.71	0.0474
I know the steps to take when my child is having an allergic reaction	1.80	1.47	0.0000	1.86	1.55	0.0000	1.87	1.45	0.0000	1.81	1.52	0.0482
I make efforts to learn new ways to help my child cope with his/her medical condition	2.38	1.61	0.0000	2.36	1.51	0.0000	2.32	1.51	0.0000	2.50	1.55	0.0000

I have a good understanding of my child's disorder	2.03	1.61	0.0000	2.09	1.62	0.0000	2.10	1.65	0.0000	2.15	1.58	0.0000
Composite Score	32.67	27.13	0.0000	33.70	28.26	0.0002	33.95	27.25	0.0000	33.43	27.41	0.0012
Items for Food Allergy-related Quality of Life–Parental Burden Scale	2.85	3.40	0.0000	3.43	4.30	0.0000	3.16	3.94	0.0000	2.65	3.01	0.0061
If you and your family were planning a holiday or vacation, how much would your choice of vacation be limited by your child's food allergy?	4.24	4.50	0.0013	4.58	5.16	0.0000	4.44	4.95	0.0000	3.70	4.14	0.0036
If you and your family were planning to go to a restaurant, how much would your choice of restaurant be limited by your child's food allergy?	3.49	3.56	0.4568	3.72	4.06	0.0015	3.64	3.91	0.0704	3.18	3.30	0.3374
If you and your family were planning to participate in social activities with others involving food, how limited would your ability to participate be because of your child's food allergy?	1.96	2.38	0.0000	2.33	3.18	0.0000	2.22	2.87	0.0000	1.90	2.49	0.0000

(continued)

Table 2.5 (continued)

	Children with peanut allergy			Children with milk allergy			Children with egg allergy			Children with tree nut allergy		
	Father	Mother	P value (father vs. mother)	Father	Mother	P value (father vs. mother)	Father	Mother	P value (fathers vs. mother)	Father	Mother	P value (father vs. mother)
In the past week, how troubled have you been by your need to spend extra time preparing meals (i.e., label reading, extra time shopping, etc.) due to your child's FA?	2.11	2.42	0.0001	2.37	3.08	0.0000	2.33	2.79	0.0000	2.05	2.56	0.0010
In the past week, how troubled have you been about your need to take special precautions before going out of the home with your child because of their food allergy?	2.01	2.54	0.0000	2.28	2.95	0.0000	2.17	2.79	0.0000	1.97	2.52	0.0000
In the past week, how troubled have you been that your child may not overcome their FA?	2.29	2.72	0.0005	2.74	3.29	0.0001	2.57	3.00	0.0040	2.30	2.54	0.1297

In the past week, how troubled have you been by the possibility of or actually leaving your child in the care of others because of their food allergy?	2.59	2.98	0.0001	2.93	3.38	0.0025	2.83	3.22	0.0128	2.48	3.02	0.0002
In the past week, how troubled have you been by frustration over other's lack of appreciation for the seriousness of food allergy?	2.53	2.94	0.0009	2.91	3.34	0.0116	2.72	3.02	0.1350	2.55	2.99	0.0245
In the past week, how troubled have you been by sadness regarding the burden your child carries because of their food allergy?	2.24	2.70	0.0006	2.55	2.98	0.0004	2.37	2.80	0.0022	2.19	2.70	0.0064
In the past week, how troubled have you been about your child's attending school, camp, daycare, or other group activity with children because of their food allergy?	2.44	3.01	0.0000	2.73	3.25	0.0005	2.61	3.11	0.0005	2.38	2.96	0.0006

(continued)

Table 2.5 (continued)

	Children with peanut allergy			Children with milk allergy			Children with egg allergy			Children with tree nut allergy		
	Father	Mother	P value (father vs. mother)	Father	Mother	P value (father vs. mother)	Father	Mother	P value (fathers vs. mother)	Father	Mother	P value (father vs. mother)
In the past week, how troubled have you been by your concerns for your child's health because of their food allergy?	2.34	2.62	0.0123	2.59	3.03	0.0006	2.50	2.84	0.0225	2.27	2.60	0.0211
In the past week, how troubled have you been with the worry that you will not be able to help your child if they have an allergic reaction to food?	2.10	2.32	0.0413	2.28	2.49	0.3938	2.20	2.33	0.5591	1.92	2.33	0.0145
In the past week, how troubled have you been with issues concerning your child being near others while eating because of their food allergy?	2.48	2.70	0.0229	2.76	3.09	0.0487	2.57	2.80	0.3487	2.29	2.55	0.0474
In the past week how troubled have you been with being frightened by the thought that your child will have a food allergic reaction?	2.32	2.74	0.0000	2.64	3.01	0.0190	2.47	2.77	0.0574	2.25	2.80	0.0008
Composite score	37.81	43.59	0.0000	42.80	50.60	0.0000	40.56	47.35	0.0000	36.13	42.21	0.0005

FA food allergy
Warren et al. (2015)

Table 2.6 Predictors of parental food allergy-related quality of life

Dependent variable: quality of life among parents of children with FA	Mother (n = 801)		Father (n = 723)	
	Coefficient (95% CI)	P value	Coefficient (95% CI)	P value
Empowerment composite score	0.008 (−0.083, 0.010)	0.862	0.002 (−0.0666, 0.071)	0.946
Food allergy severity	3.729 (0.856, 6.602)	0.011	4.476 (1.726, 7.225)	0.001
Other chronic condition	5.559 (2.418, 8.699)	0.001	1.364 (−1.553, 4.281)	0.359
Age (vs 0–1 years)				
2–5 years	−7.233 (−12.313, −2.152)	0.005	−2.282 (−7.102, 2.538)	0.353
6–10 years	−11.250 (−17.959, −4.521)	0.001	−2.926 (−9.143, 3.291)	0.356
11–13 years	−12.295 (−20.263, −4.327)	0.003	−3.082 (−11.369, 5.205)	0.465
≥14 years	5.032 (−9.587, 19.651)	0.499	−8.146 (−22.073, 5.782)	0.251
Male sex	1.139 (−1.580, 3.858)	0.411	2.697 (0.127, 5.266)	0.040
Race or ethnicity (vs other)				
Non-Hispanic white	−17.663 (−24.871, −10.456)	0.000	−7.881 (−15.039, −0.722)	0.031
Non-Hispanic black	−0.723 (−5.306, 3.860)	0.757	2.678 (−1.727, 7.082)	0.233
Hispanic	1.372 (−7.187, 9.931)	0.753	9.218 (0.905, 17.531)	0.030
Asian	−1.212 (−5.947, 3.522)	0.615	2.747 (−1.863, 7.357)	0.242
Number of siblings	−0.102 (−1.854, 1.65)	0.909	−1.437 (−3.0854, 0.212)	0.087
Health insurance (vs public)				
Private	−0.704 (−7.102, 5.694)	0.829	2.188 (−4.377, 8.753)	0.513
Other	1.725 (−9.413, 12.863)	0.761	1.805 (−8.522, 12.132)	0.732
Parental education (vs no college)				
College educated	−3.753 (−7.929, 0.422)	0.078	−0.818 (−4.756, 3.121)	0.684
Income (vs <$50,000)				
$50,000–$99,999	−1.950 (−8.286, 4.386)	0.546	1.293 (−5.265, 7.851)	0.699
≥$100,000	−1.487 (−7.798, 4.823)	0.644	−2.052 (−8.557, 4.453)	0.536
Unknown	−5.111 (−13.812, 3.590)	0.249	−3.700 (−12.413, 5.012)	0.405
Support from family and friends	−14.769 (−19.835, −9.703)	0.000	−10.145 (−15.104, −5.186)	0.000
Resources needed to care for child	−17.773 (−26.064, −9.483)	0.000	−11.458 (−21.612, −1.303)	0.027

Warren et al. (2015)

associated with significantly decreased FAQOL among mothers but not fathers. The presence of comorbid conditions resulted in no significant differences in parental empowerment. FAQOL was significantly lower for mothers and fathers of children with a history of food-induced anaphylaxis compared to parents of children with mild or moderate food allergy. Mothers of children with peanut, milk, egg, and tree nut allergies reported significantly greater empowerment and lower FAQOL than did fathers of children with these specific allergies. Parents of children with milk and egg allergies had significantly lower FAQOL scores, despite being similarly empowered to parents of children with all other allergies, even after correcting for the younger age of children with milk and egg allergies.

2.6.2   Parental Concerns and Possible Interventions to Improve Parental FAQOL

The constant risk, real or perceived, of potentially fatal anaphylaxis distinguishes food allergy from many other chronic conditions, and presents a substantial burden to parents as well as a challenge to researchers and policy-makers.

Parental concern in QOL assessments was greatest for items involving fear of allergen exposure outside the home. Elevated concern about allergic episodes occurring outside the home is in keeping with the relationship between FAQOL and empowerment presented here; in the case of food allergy, being aware of risks may lead to greater stress. Given the prevalence of poor QOL among parents of children with food allergy, there is a need for comprehensive means of addressing low FAQOL in this population. Ideally, knowing more about food allergy would mean a higher quality of life, however there are too many variables to be able to expect this sort of outcome in the near term for all parents. Similarly, in best cases, empowerment would positively correlate with FAQOL outcomes.

The differences seen between maternal and paternal QOL among parents of children with food allergy have also been demonstrated in asthma (Hederos et al. 2007), developmental disorders (Yamada et al. 2012), and other chronic conditions (Goldbeck 2006) as well. The cause of this difference between mothers and fathers needs further elucidation. Our findings support the importance of social networks for parents of food-allergic children, especially mothers. Building mutually supportive social networks may increase parental FAQOL (Warren et al. 2015).

2.7   Food Allergen Labeling and Purchasing Habits in the U.S. and Canada

Mandatory labeling for common food allergens exists in the U.S., with a similar set of guidelines in Canada. Standardization of labeling practices has increased safety for consumers with food allergies. However, precautionary labeling (e.g., "may contain," "processed in a facility"), which is neither standardized nor regulated, has become more common, and its effects on the purchasing and eating habits of individuals with food allergy are largely unknown. We therefore sought to characterize the purchasing habits of a cohort of 6684 people with food allergy or their caretakers, from the U.S. and Canada (Marchisotto et al. 2016).

2.7.1   Labeling Knowledge and Purchasing Behavior

The results showed that 29% of respondents were unaware of laws requiring labeling of major allergens. Moreover, 46% of respondents were unaware of or mistaken about guidelines regarding advisory or precautionary labeling.

Purchasing habits were evaluated by type of labeling, severity of allergy, and country of purchase. Twelve percent of respondents reported purchasing food labeled to indicate that it "may contain" their allergen, whereas 40% reported purchasing foods labeled "manufactured in a facility that also processes" their allergen (Marchisotto et al. 2016).

2.7.2   Implications of Labeling Practices and Purchasing Behavior

None of the precautionary labels evaluated have specific legal definitions, leaving consumers without safety information. The potential danger of non-standardized precautionary labeling is especially salient for children and families with limited access to fresh food and greater dependence on packaged goods. This observation may influence the differences in food allergy outcomes seen among children of different socioeconomic statuses (Marchisotto et al. 2016). Children of families who predominantly have access to packaged foods may be more likely to suffer due to unclear labeling, simply as a product of their consuming more of these foods in their diets.

2.8   The Economic Impact of Childhood Food Allergy in the U.S.

Attempts to assess the economic burden of food allergy in the U.S. have been limited by the use of federal diagnosis code data, which may fail to capture all cases and does not take into consideration direct and indirect costs to families. We used a cross-sectional survey of parents and caregivers of children with food allergy ($n = 1643$) to quantify the costs associated with caring for a child with food allergy (Gupta et al. 2013). Costs considered included direct medical expenses (e.g., physician fees and fees for ED visits and hospitalizations), out-of-pocket expenses (e.g., safe foods, clinic or ED copays, medications), lost-labor costs (e.g., taking time off work for allergy-related reasons), and opportunity costs (e.g., needing to leave or change jobs). Caregivers were also questioned about their willingness to pay for effective food allergy treatments. We assessed the distribution of these costs to families across multiple demographic categories. The total cost of food allergy in the U.S. was estimated to be $24.8 billion annually (Gupta et al. 2013).

2.8.1 Direct Medical Costs of Childhood Food Allergy

Direct medical costs include costs of the health care system for the diagnosis, treatment, and prevention of childhood food allergy. Data were collected by asking caregivers of children with food allergy about outpatient, ED, and inpatient visits over a 1-year span. Costs for each type of encounter were estimated from several sources, including Medicare data, the Healthcare Cost and Utilization Project Nationwide Emergency Department Sample, and the Healthcare Cost and Utilization Project Nationwide Inpatient Sample.

Direct medical costs were estimated to be $4.3 billion per year (Table 2.7). Hospitalizations accounted for almost 50% of this figure, while ED visits made up 18% of annual direct medical costs. Specialist visits of other sorts made up the remainder of direct medical expenses.

2.8.2 Costs Borne by Families

Costs borne by families included out-of-pocket, lost labor productivity, and opportunity costs. The total cost borne by caregivers was $20.5 billion per year.

Table 2.7 Direct medical costs of childhood food allergy

Characteristic	Children with visit, % (SE)	Visits per child, mean (SE)	Cost, US$		Overall annual (in millions)
			Visit	Child	
Visits					
Pediatrician	42 (2)	0.82 (0.05)	112	92	543
Allergist	41 (2)	0.79 (0.05)	175	138	819
Pulmonologist	14 (1)	0.07 (0.01)	175	12	71
Nutritionist	17 (1)	0.16 (0.04)	100	16	96
Alternative provider	17 (1)	0.23 (0.05)	100	23	136
Emergency department	13 (1)	0.18 (0.02)	711	129	764
Inpatient hospitalization stays	4 (1)	0.05 (0.01)	6269	314	1863
Total direct medical costs				724	4292

Gupta et al. (2013)

Table 2.8 Out-of-pocket costs of childhood food allergy

Variable	% Reporting cost (SE)	Mean direct out-of-pocket costs, US$ (SE)	Cost per child, US$	Overall annual cost (in millions), US$
Visits to the physician's office or health clinic (including copays)	52.5 (2.2)	160 (14)	84	499
Visits to the emergency room (including copays)	16.1 (1.6)	247 (42)	40	235
Overnight stays at the hospital	10 (1.4)	411 (182)	41	244
Travel to and from health care visits (including ambulance use; parking expenses)	27.7 (1.8)	91 (14)	25	149
Epinephrine injectors (Epipen, Epipen Jr.)	35.9 (1.9)	87 (4)	31	184
Antihistamines (Allegra, Benadryl, Claritin, Zyrtec)	50.8 (2.2)	62 (4)	32	188
Other prescription/nonprescription medication	29.3 (1.9)	122 (13)	36	211
Non-traditional medicine (such as herbal products)	15 (1.6)	123 (30)	19	110
Costs associated with special diets and allergen-free foods	37.7 (2.0)	756 (59)	285	1689
Additional/change in child care	6.7 (0.8)	2158 (323)	145	857
Legal guidance	2.3 (0.6)	402 (122)	9	55
Counseling or mental health services	4.5 (0.7)	571 (123)	26	152
Special summer camp	3 (0.7)	702 (183)	21	125
A change in schools was needed due to this child's food allergy	4.2 (0.7)	2611 (497)	110	650
Other out-of-pocket expenses (e.g., cleaning supplies, skin care products, transportation)	9.2 (1.1)	396 (86)	36	216
Any out-of-pocket costs	74.3 (2.1)	1252 (90)	931	5516

Gupta et al. (2013)

2.8.3   Out-of-Pocket Costs

The total out-of-pocket cost to families was $5.5 billion per year (Table 2.8). Allergen-free foods were the greatest out-of-pocket expense, accounting for over 30% of the annual figure. Changes in child care and schooling together comprised 27% of all out-of-pocket costs.

2.8.4   Lost Labor Costs to Families

Lost labor was defined as the sum of unearned wages due to caregiver hours spent on food allergy-related healthcare visits, based on the mean national hourly labor wage. Lost labor costs totaled $773 million (Gupta et al. 2013).

Table 2.9 Opportunity cost of childhood food allergy

Characteristic	Reporting, % (SE)	Opportunity, mean (SE)	Cost, US$ per child	Overall annual (in billions)
Choice of career has been restricted	5.7 (0.9)	15 655 (2471)	892	5.3
A job had to be given up	4.9 (0.7)	29 657 (4151)	1453	8.6
A job was lost through dismissal	1.9 (0.6)	14 849 (7479)	282	1.7
A job change was required	2.5 (0.6)	10 605 (3161)	265	1.6
Any job-related opportunity cost (total amount)	9.1 (1.0)	32 719 (4166)	2977	17.6
Any job-related opportunity cost (maximum amount)	9.1 (1.0)	26 363 (2545)	2399	14.2

Gupta et al. (2013)

2.8.5 Opportunity Costs to Families

A job-related opportunity cost was reported by 9.1% of caregivers. These costs consisted of restricted career choice, need to leave a job, need to change the jobs, or losing a job. The opportunity cost to families was calculated as the product of the percent of caregivers reporting lost opportunity in the labor market, the mean reported cost of these opportunities, and the number of children with food allergy in the U.S. The annual lost opportunity cost of forgone labor was estimated at $14 billion per year (Table 2.9).

2.8.6 Willingness to Pay

Willingness to pay (WTP) was calculated by surveying caregivers about the amount they would be willing to spend on a hypothetical treatment for food allergy that would allow their child to eat all foods. Annual WTP equaled $20.8 billion per year. This cost was not the same as their current medical costs, but instead asked about an imaginary treatment which would enable consumption of all foods without causing an allergic reaction.

2.8.7 What the Economic Impact of Childhood Food Allergy in the U.S. Tells Us

To our knowledge, this was the first study to comprehensively quantify the economic impact of food allergy in the U.S. We have shown that families affected by food allergy bear a substantial financial burden. The total cost of food allergy was estimated to be $24.8 billion annually. Over 80% of this cost is borne by families of those with allergies. To put these numbers in context, we compare them with those of asthma, as asthma affects a similar number of children in the U.S. (Fox et al. 2013). Direct medical costs for asthma have been estimated at $3259 per person (regardless of age) per year, which is roughly five times the cost per food-allergic child ($724 per child) (Barnett and Nurmagambetov 2011). Prescription medications account for over 50% of the direct medical costs associated with asthma (Fox et al. 2013). In comparison, few prescriptions exist to treat food allergy. Instead, food allergy requires families to make accommodations in many other areas of life, each with substantial financial costs to the family. Unlike prescription medicine, the opportunity cost of caregiver time is seldom covered by insurance. These sorts of expenses are especially burdensome to low-income families, who cannot easily afford special foods or to sacrifice hours at work.

Annual WTP was estimated at $20.8 billion per year. This figure is similar to the reported annual cost borne by caregivers ($20.5 billion) and therefore seems to confirm the validity and consistency of the two analytic approaches employed in this study. That is, the fact that WTP closely matches the amount actually spent by families supports the idea that the method of measuring WTP is a successful model (Gupta et al. 2013).

2.9   Socioeconomic Disparities in the Economic Impact of Childhood Food Allergy in the U.S.

Little is known about how food allergy-related costs are distributed across racial/ethnic and socioeconomic lines. We therefore set out to quantify disparities in the distribution of direct and out-of-pocket costs associated with childhood food allergy between socioeconomic and racial/ethnic groups (Bilaver et al. 2016). Data were obtained via cross-sectional survey of 1643 U.S. caregivers with a food-allergic child. Variables surveyed were direct medical costs (calculated from parent-reported outpatient visits), ED visits and hospitalizations, and out-of-pocket costs (calculated from parent-reported amounts spent on medication, insurance co-pays, mental health services, legal services, special schooling, child care, camp tuition, and special food). Five racial (African American, Asian, Hispanic, White, Other) and three socioeconomic (household income of: <$50K, $50–$99K, ≥$100K) groups were compared (Bilaver et al. 2016).

2.9.1   Examining Medical Access by Socioeconomic Status (SES) and Race/Ethnicity

We have previously shown that African American and Hispanic children are significantly less likely to have formal diagnoses of food allergy (Sect. 2.2.1.3). However, there has been no reported evidence of a relationship between race or ethnicity and food allergy severity. It is also known that children from higher income households are more likely than children from low-income households to be prescribed epinephrine auto-injectors (Coombs et al. 2011). Higher income families are also more likely to have administered epinephrine to their children before reaching the emergency department in cases of food-induced anaphylaxis (Huang et al. 2012). These higher rates of prescription and of epinephrine use are not indicative of children from high-income families; we have shown that children with public insurance have higher yearly rates of ED visits for food-induced anaphylaxis than children with private insurance. Instead, wealthier families may have an easier time accessing healthcare resources.

2.9.1.1   Medical Costs by SES

Households making <$50K spent 2.5 times the amount that higher-income families spent on ED visits and hospitalizations due to food allergy. Conversely, spending on specialists was significantly lower in the lowest income group than the highest income group. Families in the highest income group also spent significantly more (>2 times more) on medication than did the lowest income group (Table 2.10).

Table 2.10 Direct and out-of-pocket mean annual costs by socioeconomic status (household income in US$) (top) and by race/ethnicity (bottom)

Direct and out-of-pocket mean annual costs (SE), US$ by household income			
Type of cost	**<$50K**	**$50K–99K**	**≥$100K**
Total Direct Costs borne by health care system	1374 (274)	1024 (125)	940 (128)
ER and Hospitalization costs*	1021 (209)	434 (106)	416 (94)
Specialist costs**	228 (21)	330 (27)	311 (18)
Total Out-of-Pocket Costs borne by families	3174 (858)	3434 (658)	5062 (1168)
Medication costs***	171 (26)	275 (30)	366 (44)
Special food costs	744 (216)	941 (230)	1545 (347)

$*p < 0.05$, $**p < 0.01$, $***p < 0.001$ for F-test of equality of means across groups

Bilaver et al. (2016)

Type of cost	**White**	**African American**	**Hispanic**	**Asian**
Total Direct Costs borne by health care system***	999 (104)	493 (109)	643 (224)	885 (514)
ER and Hospitalization costs***	504 (79)	108 (60)	395 (220)	1271 (630)
Specialist costs***	310 (13)	157 (40)	127 (37)	101 (36)
Total Out-of-Pocket Costs borne by families***	4203 (750)	395 (452)	1093 (856)	1327 (1948)
Medication costs***	312 (28)	52 (18)	148 (78)	87 (37)
Special food costs***	1213 (200)	177 (501)	219 (281)	148 (290)

$*p < 0.05$, $**p < 0.01$, $***p < 0.001$ for F-test of equality of means across groups

Bilaver et al. (2016)

2.9.1.2   Medical Costs by Race/Ethnicity

Significant differences in spending were also seen across racial/ethnic lines. An F-test for equality of means across the five racial/ethnic groups, measuring total direct costs borne by the healthcare system, returned a p-value less than 0.001. The same level of significance was seen between groups for equality of means when examining total out-of-pocket costs. White families spent more on specialists, medications, and special foods than did families from other racial or ethnic groups. African American children incurred the lowest direct medical expenses and spent the least on out-of-pocket costs per year of all racial and ethnic groups (Table 2.10).

2.9.1.3   What Socioeconomic Disparities in the Economic Impact of Childhood Food Allergy Tell Us

Given these findings, we hypothesize that children from low income and non-White families may have reduced access to preventative health care and subspecialty visits, and often rely on emergency treatment for the acute adverse reactions. Increasing access to subspecialist care, safe foods, and epinephrine is therefore crucial. Epinephrine availability in public places and equity in the management of food allergy are issues that still needs more effort (Bilaver et al. 2016).

2.10   Asthma and Food Allergy Management and Emergency Epinephrine Use for Food Allergy Reactions in Chicago Public Schools

The provision of stock or undesignated epinephrine auto-injectors (EAIs) in schools and other public places is the subject of active investigation and debate. During the 2012–2013 school year, Chicago Publics Schools (CPS) began providing undesignated epinephrine to district schools. We collected data on epinephrine auto-injector use at the end of the school year and analyzed them to elucidate effects on food-allergy outcomes (Gupta et al. 2014). To our knowledge, these studies are the first to examine asthma and food allergy reporting and management, as well as the impact of providing stock epinephrine auto-injectors on food allergy outcomes, in a large urban school district.

2.10.1   Distribution of EAI Use in CPS

Thirty-eight district-issued EAIs were administered during the first year of the program. The majority of these injections were given in elementary schools. EAIs were most commonly used on Chicago's North-Northwest side (37%); however nearly three quarters as many EAIs were used in schools on Chicago's Far South Side as were used on the North-Northwest side. The overwhelming discrepancy in diagnosis of food allergies between these two parts of the city (47% versus 8%) is not reflected in the usage of EAIs. More than half of all district-issued EAIs were used for the recipient's first-time anaphylactic event. Food was identified as the trigger in more than half of all reactions, while the cause of over one-third of reactions was unknown (Table 2.11).

Table 2.11 Characteristics and triggers of the administer of district-issued epinephrine

Variable	Frequency, % (n)
Person with allergic reaction	
Chicago public school student	92.1 (35)
Chicago public school staff	7.9 (3)
School type	
Elementary school	63.2 (24)
High school	36.8 (14)
Geographic collaborative	
Far South Side	26.3 (10)
North-Northwest	36.8 (14)
South Side	13.2 (5)
Southwest Side	10.5 (4)
West Side	13.2 (5)
First-time incident	
Yes	55.0 (21)
No	45.0 (17)
Administered by	
School nurse	76.3 (29)
Other school staff	18.4 (7)
Self-administered	5.3 (2)
Epinephrine dose	
Adult	57.9 (22)
Junior	23.7 (9)
Missing	18.4 (7)
911 Called	
Yes	81.6 (31)
No	2.6 (1)
Missing	15.8 (6)
Allergic trigger—food	55.3 (21)
Peanut	18.4 (7)
Tree nut	2.6 (1)
Shellfish	2.6 (1)
Fin fish	13.2 (5)
Food—other	13.2 (5)
Food—unknown	5.3 (2)
Allergic trigger—other	10.5 (4)
Insect venom	5.3 (2)
Animals	2.6 (1)
Grass	2.6 (1)
Allergic trigger—unknown	34.2 (13)

DeSantiago-Cardenas et al. (2015)

Table 2.12 Adjusted odds ratios of asthma and food allergy diagnoses among CPS students

Variable	Asthma	Food allergy
Age (vs 3–5 years)		
6–10	10.4 (5.7–19.2)**	1.3 (1.2–1.4)**
11–13	12.9 (7.0–23.8)**	1.0 (0.9–1.1)
$14	11.0 (6.0–20.2)**	0.7 (0.6–0.8)**
Gender		
Male versus female	1.4 (1.4–1.5)**	1.3 (1.2–1.4)**
Race/ethnicity (vs white)		
Black	2.3 (2.2–2.4)**	1.1 (1.0–1.3)*
Hispanic	1.3 (1.2–1.4)**	0.8 (0.7–0.9)**
Asian	0.8 (0.7–0.9)**	1.4 (1.2–1.6)**
Multiracial all	1.7 (1.5–2.0)**	1.5 (1.3–1.9)**
Other	1.1 (0.9–1.3)	1.0 (0.7–1.3)
Free/reduced lunch program (vs did not apply)		
Free	1.1 (0.9–1.1)	0.3 (0.3–0.4)**
Reduced	1.0 (0.9–1.0)	0.5 (0.4–0.5)**
45-day temporary free	1.0 (0.8–1.3)	0.5 (0.4–0.8)**
Denied	1.1 (1.0–1.2)*	0.8 (0.8–0.9)**
Geographic region (vs North-Northwest Side)		
Far South Side	0.7 (0.7–0.8)**	0.7 (0.6–0.7)**
South Side	0.7 (0.7–0.8)**	0.6 (0.6–0.7)**
Southwest Side	0.7 (0.7–0.8)**	0.5 (0.5–0.6)**
West Side	0.9 (0.9–0.9)**	0.6 (0.6–0.7)**

Data are presented as OR (95% CI). *$p < 0.05$, **$p < 0.001$
Gupta et al. (2014)

2.10.2   CPS Food Allergy Data by Race, SES, and Geographic Region

Odds of having food allergy were significantly higher among African American (OR = 1.1) and Asian students (OR = 1.4) than White students. African American and Hispanic students made up substantial portions of all food-allergic students (prevalence of food allergy and asthma is detailed in Table 2.12).

Odds of having food allergy were significantly lower among students who received free (OR = 0.3) or reduced (OR = 0.5) lunch than among those who did not. Odds of food allergy were also significantly lower among all other geographic regions when compared with the city's North-Northwest Side.

2.10.3   Emergency Action Plan Data in CPS

A 504 plan is an emergency action plan put together by parents and schools to ensure appropriate steps are taken in the event of anaphylaxis or an acute asthma exacerbation. Fifty-one percent of CPS students with food allergy had a 504 plan on file for the aforementioned school year. Among students with food allergy and/or asthma, odds of having a 504 Plan decreased as age increased. Overall odds of having a 504 plan were lower for boys than girls, African American and Hispanic students than White students, students who qualified for free or reduced lunch than those who did not, and among students from all other areas of Chicago when compared with the North-Northwest Side. Similar odds of having a 504 plan were found between students who had food allergy versus students who had asthma. Forty percent of all food-allergic students also had asthma, while only 9% of students with asthma also had food allergy. Whether this disparity is due to variability in disease prevalence or differences in reporting of chronic conditions to CPS is uncertain (Prevalence of 504 plans is detailed in Table 2.13).

Table 2.13 Adjusted odds ratios of having school health management plans (504 plans) among CPS students, by asthma and food allergy diagnosis

Variable	Asthma	Food allergy
Age (vs 3–5 years)		
6–10	1.2 (1.0–1.3)*	1.1 (0.9–1.3)
11–13	0.8 (0.7–0.8)**	0.7 (0.5–0.9)*
$14	0.5 (0.4–0.6)**	0.5 (0.4–0.6)**
Gender		
Male versus female	0.8 (0.7–0.8)**	0.7 (0.7–0.9)*
Race/ethnicity (vs white)		
Black	0.5 (0.4–0.6)**	0.6 (0.5–0.7)**
Hispanic	0.8 (0.7–0.9)*	0.9 (0.7–1.2)
Asian	1.0 (0.8–1.3)	0.9 (0.6–1.2)
Multiracial all	1.0 (0.8–1.4)	1.0 (0.7–1.6)
Oother	0.9 (0.6–1.4)	0.8 (0.4–1.5)
Free/reduced lunch program (vs did not apply)		
Free	0.6 (0.5–0.7)**	0.5 (0.4–0.6)**
Reduced	0.8 (0.6–0.9)*	0.7 (0.5–0.9)*
45-day temporary free	0.7 (0.4–1.2)	0.4 (0.2–0.9)*
Denied	1.0 (0.8–1.2)	0.9 (0.7–1.1)
Immunization status		
Compliant versus not compliant	1.4 (1.1–1.8)*	1.7 (1.1–2.7)*
Geographic region (vs North-Northwest Side)		
Far South Side	0.6 (0.5–0.7)**	0.4 (0.3–0.6)**
South Side	0.6 (0.5–0.6)**	0.5 (0.4–0.7)**
Southwest Side	0.6 (0.5–0.6)**	0.5 (0.4–0.6)**
West Side	0.5 (0.5–0.6)**	0.6 (0.5–0.7)**
Chronic condition		
Asthma	—	1.9 (1.7–2.2)**
Food allergy	4.1 (3.7–4.6)**	—

Data are presented as OR (95% CI). *$p < 0.05$, **$p < 0.001$
Gupta et al. (2014)

2.10.4   Conclusions and Recommendations based on EAI and Emergency Action Plan Data from CPS

CPS is the largest school district in the country to have taken on the challenge of providing its schools with undesignated EAI for the emergency treatment of anaphylaxis. We have described a discrepancy between documented food allergy and EAI usage in CPS. These findings reinforce previous evidence of underdiagnosis of food allergy among low income and minority populations, and therefore the need for undesignated EAIs. Research has shown that 20–25% of children experience their first reaction in schools. Moreover, 30% of nurses report that they have had to use one child's EAI on another. These facts, along with those already presented, make the necessity of undesignated EAIs and 504 plans clear.

Our data show that 504 plans are substantially underutilized among CPS students. Underutilization of management plans has been reported widely in the recent literature, suggesting it is a problem across school districts. Moreover, although African American and Hispanic students were disproportionately affected by asthma and food allergy, they were significantly less likely than white CPS students to have school health management plans on file. Low-income students (approximated as those who participated in the free/reduced lunch program) were also significantly less likely than more affluent students to have 504 plans on file. It is unclear whether racial and socioeconomic variability in food allergy is intrinsic to the disease process; however, it is possible that the disparities seen here are attributable, at least in part, to barriers to care and diagnosis among racial/ethnic minorities and economically disadvantaged households. Given the uneven geographic distribution of healthcare providers throughout Chicago and the difficulty that low-income families face in accessing healthcare resources, it is unsurprising that the diagnosis of food allergies does not mirror, its incidence.

Given that this study was performed using CPS data and therefore only includes students with physician-verified food allergies and asthma, rates of these conditions may be higher than reported among the low-income populations within CPS. Currently, a CPS student needs a physician's verification to create a 504 plan. Low-income students, who often depend on school support programs, may therefore be excluded due to limited access to healthcare resources. A potential low-cost solution would be to offer 504 plans without requiring physician verification of a food allergy diagnosis (Gupta et al. 2014).

2.11   Conclusion

We have detailed the prevalence of food allergy, its geographic distribution, and the burden it places on the medical system, families, minority and low-income communities, and schools. Under-privileged populations bear a disproportionate share of this burden. This review of our research provides a big-picture of the public health impact of food allergy. Given the economic, social, and psychological burdens that food allergy presents with, more systematic solutions are needed. These costs will eventually accumulate along with the increasing rates of food allergy diagnosis. While progress toward a cure is incremental, broad access to food allergy education and undesignated epinephrine can be implemented now. An understanding of best practices and the emergency care of food-allergic reactions are tools currently available. It is vital that we take strides to distribute these lifesaving resources.

References

Barnett, S.B.L., and T.A. Nurmagambetov. 2011. Costs of asthma in the United States: 2002-2007. *Journal of Allergy and Clinical Immunology* 127: 145–152.

Ben-Shoshan, M., D.W. Harrington, L. Soller, J. Fragapane, L. Joseph, Y. St Pierre, S.B. Godefroy, S.J. Elliot, and A.E. Clarke. 2010. A population-based study on peanut, tree nut, fish, shellfish, and sesame allergy prevalence in Canada. *Journal of Allergy and Clinical Immunology.* 125: 1327–1335.

Bilaver, L.A., K.M. Kester, B.M. Smith, and R.S. Gupta. 2016. Socioeconomic disparities in the economic impact of childhood food allergy. *Pediatrics.* 137: e20153678.

Cohen, B.L., S. Noone, A. Muñoz-Furlong, and S.H. Sicherer. 2004. Development of a questionnaire to measure quality of life in families with a child with food allergy. *Journal of Allergy and Clinical Immunology.* 114: 1159–1163.

Coombs, R., E. Simons, R.G. Foty, D.M. Stieb, and S.D. Dell. 2011. Socioeconomic factors and epinephrine prescription in children with peanut allergy. *Paediatrics and Child Health.* 16: 341–344.

DeSantiago-Cardenas, L., V. Rivkina, S.A. Whyte, B.C. Harvey-Gintoft, B.J. Bunning, and R.S. Gupta. 2015. Emergency epinephrine use for food allergy reactions in Chicago public schools. *American Journal of Preventive Medicine.* 48: 170–173.

Dyer, A.A., C.H. Lau, T.L. Smith, B.M. Smith, and R.S. Gupta. 2015. Pediatric emergency department visits and hospitalizations due to food-induced anaphylaxis in Illinois. *Annals of Allergy, Asthma and Immunology.* 115: 56–62.

Fox, M., M. Mugford, J. Voordouw, J. Cornelisse-Vermaat, G. Antonides, B. de la Hoz Caballer, I. Cerecedo, J. Zamora, E. Rokicka, M. Jewczak, A.B. Clark, M.L. Kowalski, N. Papadopoulos, A.C. Knulst, S. Seneviratne, S. Belohlavkova, R. Asero, F. de Blay, A. Purohit, M. Clausen, B. Flokstra de Blok, A.E. Dubois, M. Fernandez-Rivas, P. Burney, L.J. Frewer, and C.E. Mills. 2013. Health sector costs of self-reported food allergy in Europe: A patient-based cost of illness study. *The European Journal of Public Health.* 23: 757–762.

Goldbeck, L. 2006. The impact of newly diagnosed chronic paediatric conditions on parental quality of life. *Quality of Life Research.* 15: 1121–1131.

Gupta, R.S., E.E. Springston, M.R. Warrier, B. Smith, R. Kumar, J. Pongracic, and J.L. Holl. 2011. The prevalence, severity, and distribution of childhood food allergy in the United States. *Pediatrics.* 128: e9–e17.

Gupta, R.S., E.E. Springston, B. Smith, M.R. Warrier, J. Pongracic, and J.L. Holl. 2012. Geographic variability of childhood food allergy in the United States. *Clinical Pediatrics.* 51: 856–861.

Gupta, R., D. Holdford, L. Bilaver, A. Dyer, J.L. Holl, and D. Meltzer. 2013. The economic impact of childhood food allergy in the United States. *JAMA Pediatrics.* 167: 1026–1031.

Gupta, R.S., V. Rivkina, L. DeSantiago-Cardenas, B. Smith, B. Harvey-Gintoft, and S.A. Whyte. 2014. Asthma and food allergy management in Chicago public schools. *Pediatrics.* 134: 729–736.

Gupta, R., M.M. Walkner, C. Lau, D. Caruso, X. Wang, J.A. Pongracic, and B. Smith. 2015. Food allergy sensitization and presentation in siblings of food allergic children. *Annals of Allergy Asthma and Immunology.* 115: A27–A27.

Hederos, C.A., S. Janson, and G. Hedlin. 2007. A gender perspective on parents' answers to a questionnaire on children's asthma. *Respiratory medicine.* 101: 554–560.

Huang, F., K. Chawla, K.M. Järvinen, and A. Nowak-Węgrzyn. 2012. Anaphylaxis in a New York City pediatric emergency department: Triggers, treatments, and outcomes. *Journal of Allergy and Clinical Immunology.* 129: 162–168.

Koren, P.E., N. DeChillo, and B.J. Friesen. 1992. Measuring empowerment in families whose children have emotional disabilities: A brief questionnaire. *Rehabilitation Psychology.* 37: 305.

Marchisotto, M.J., L. Harada, O. Kamdar, B. Smith, K. Khan, S.H. Sicherer, S.L. Taylor, V. LaFemina, M.A. Muraro, S. Waserman, and R. Gupta. 2016. Food allergen labeling and purchasing habits in the US and Canada. *Journal of Allergy and Clinical Immunology.* 137: AB81.

Warren, C.M., R.S. Gupta, M.W. Sohn, E.H. Oh, N. Lal, C.F. Garfield, D. Caruso, X. Wang, and J.A. Pongracic. 2015. Differences in empowerment and quality of life among parents of children with food allergy. *Annals of Allergy, Asthma and Immunology.* 114: 117–125.

Yamada, A., M. Kato, M. Suzuki, M. Suzuki, N. Watanabe, T. Akechi, and T.A. Furukawa. 2012. Quality of life of parents raising children with pervasive developmental disorders. *BMC Psychiatry.* 12: 1.

Chapter 3
Current U.S. State Legislation Related to Food Allergen Management

Wendy Bedale

3.1 Introduction

Effective management of food allergens and allergic reactions are important goals for restaurants, schools, day care facilities, higher education institutions, and many other establishments where food may be consumed. Strategies to prevent unwanted food allergen exposure and to effectively treat anaphylactic reactions resulting from accidental exposures are important cornerstones of food allergen/food allergy management. Increased awareness and knowledge of food allergens, particularly for food handlers, is an important way to prevent accidental exposure (or errors in ingredient disclosure), especially since many serious anaphylaxis events occur in restaurants or outside the home (Eigenmann and Zamora 2002; Weiss and Munoz-Furlong 2008). Prompt treatment of anaphylaxis with epinephrine can save lives (Xu et al. 2014) and reduce hospitalizations (Fleming et al. 2015). This efficacy together with its relative safety (Manivannan et al. 2014) has prompted legislative initiatives to facilitate availability of intramuscular (autoinjector) epinephrine in schools and other public locations where food allergy reactions can occur.

Because food allergies may be categorized in the general public's minds with food intolerances (Yee Ming and Hui 2015) or voluntary food restrictions made as lifestyle choices, the seriousness of food allergy reactions is not always appreciated by those who may serve food to a food allergic individual (Gupta et al. 2009). Increasing awareness of food allergies can prevent accidental allergen exposures by making sure those responsible for preparing and serving food know how to prevent cross-contact and accidental exposures. An understanding of the consequences of food allergies and their management can also facilitate rapid and effective intervention in the event of an exposure.

W. Bedale (✉)
Food Research Institute, University of Wisconsin-Madison, Madison, WI 53706, USA
e-mail: bedale@wisc.edu

© Springer International Publishing AG 2018   55
T.-J. Fu et al. (eds.), *Food Allergens*, Food Microbiology and Food Safety,
DOI 10.1007/978-3-319-66586-3_3

To improve food allergy knowledge or management, voluntary measures such as guidelines can achieve success, but using the power of the law may be more effective. The Canadian province of Ontario was an early (2005) adopter of legislation ("Sabrina's Law") to improve management of anaphylactic events at schools (Cicutto et al. 2012). In 2003, Sabrina Shannon died from an anaphylactic reaction that occurred while she was in school (Gregory 2012). In her honor, Sabrina's Law required school boards to develop anaphylaxis management plans and to provide training to school staff regarding epinephrine administration (Food Allergy Canada 2017). A recent study compared schools within Ontario to schools in other Canadian provinces with voluntary policies, concluding that schools in legislated environments were more effective in helping students at risk for anaphylaxis (Cicutto et al. 2012).

Laws can, therefore, have an impact on food allergy/allergen management. Such laws can take the form of legal requirements or mandates for certain actions which improve control of food allergens or facilitate rapid and effective treatment when an accidental exposure occurs. Another approach is to provide incentives for these types of actions. Because health- and education-related laws are typically made at the state or local level, the federal government may use "carrot" rather than "stick" approaches to encourage state and local governments to pass laws to achieve certain goals. U.S. federal law has been used to encourage state and local efforts in allergen management.

This chapter reviews U.S. state legislation aimed to improve food allergy/allergen management. Background on key federal laws and other initiatives related to food allergy/allergen management that have influenced state legislation is provided. As can be seen in Fig. 3.1, many of these federal initiatives occurred between 2000 and 2013, which corresponds to when certain key state initiatives germinated.

The role of food allergy tragedies such as Sabrina Shannon's in shaping U.S. state efforts is discussed, after which various state-initiated legislation in areas of food allergen management in restaurants, schools, colleges and universities, and public entities is described. Current and potential future state initiatives related to food allergens are briefly discussed. The reader is also referred to Chap. 4, which presents an overview of FARE's advocacy efforts related to national and state legislation to improve food allergy awareness, management, and treatment.

3.2 Background on Federal Laws, Guidelines, and Initiatives Related to Food Allergen Management

Federal laws have helped set the stage for state and local efforts related to food allergy/allergen management in a variety of ways, especially by increasing awareness of food allergies and providing examples or guidelines for states to use to create their own requirements.

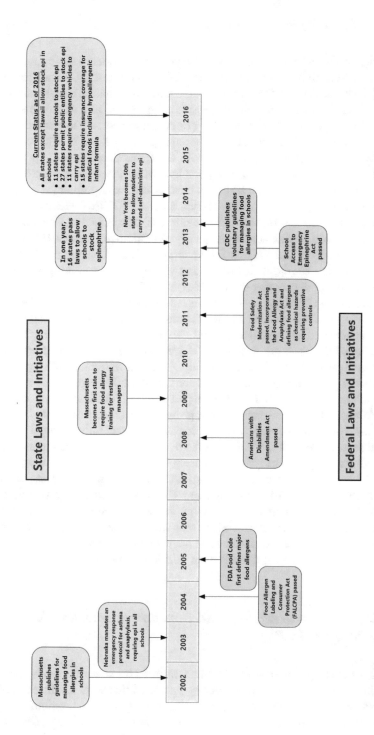

Fig. 3.1 Federal and state laws and initiatives related to food allergy/allergen management. Sources: Sheetz et al. (2004), Murphy et al. (2006), Collins (2014), FDA (2005), Terry (2014)

3.2.1   Food Allergy as a Disability

Food allergies, even severe ones, have not always been considered disabilities.

Two U.S. federal civil rights laws, the Rehabilitation Act of 1973 and the 1990 American Disabilities Act (ADA), prevent discrimination against individuals with physical or mental impairments. Such discrimination is defined as anything that prevents a disabled person from participating or benefiting equally to a nondisabled person, with no additional requirements or surcharges (Lynch 2017). Both of these laws prohibit this type of discrimination, with the Rehabilitation Act applying to programs and activities that are federally funded, while the more comprehensive ADA covers private employers, all state and local government programs (including public schools), and places of public accommodation (O'Brien-Heinzen 2010). Initially under these laws, a narrow interpretation of "disabled" was applied, and food allergic individuals did not often qualify (O'Brien-Heinzen 2010). Courts ruled in cases from 1999 to 2008 that food allergic individuals were not "substantially limited" in their eating, their impairment was episodic, and mitigating measures (epinephrine) were considered available (Roses 2011).

The Americans with Disabilities Amendment Act of 2008 greatly expanded the interpretation of "disabled," eliminating the restrictions previously used to exclude food allergic individuals from coverage (Roses 2011). To illustrate how this law has changed how food allergic individuals are viewed, the U.S. Department of Education now even uses allergy as an example of a hidden disability. A student with a food allergy is now able to obtain a written management program for their disability under Section 504 of the Rehabilitation Act of 1973 if they attend a public school (or some private schools) (O'Brien-Heinzen 2010). The management program documents and puts in place formal measures to ensure the student can participate safely and fully in school.

The legal recognition that food allergies are disabling occurred at about the same time that state laws began to be written requiring food allergy/allergen management in schools and restaurants and may have contributed to the overall awareness of food allergies and the importance of managing them. Defining a food allergy as a disability does not mean that restaurants or other public accommodations have to serve allergen-free food; however, such establishments should be able to provide information about the ingredients of menu items and be prepared to omit allergenic ingredients from dishes upon request if it is not overly difficult to do so (Lynch 2017).

3.2.2   Food Allergen Labeling and Consumer Protection Act of 2004 (FALCPA)

In 1999, FDA found that one-quarter of all manufactured baked goods, candy, and ice cream products tested contained peanut or egg proteins that were not listed on the product label (FDA 2004). The need to have better labeling to prevent serious allergic reactions was apparent. In response, a 2004 amendment to the Federal Food, Drug, and Cosmetic Act known as the Food Allergen Labeling and Protection Act (FALCPA) was

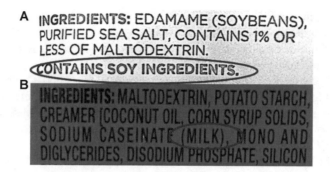

Fig. 3.2 Food ingredient labels declaring major allergens

created, which required packaged foods to indicate the presence of one of the eight major allergens (wheat, soy, milk, eggs, peanuts, tree nuts, fish, and shellfish) on the product label. The allergen may be listed in one of several ways on the label, either after the ingredient list (following the word "Contains," as in the circled portion of Fig. 3.2a) or within the ingredient list by following the allergen-containing ingredient with the name of the allergen in parentheses (Fig. 3.2b). Note the label in Fig. 3.2a actually contains both types of allergen labels, which should make it very clear to a consumer that the common name "edamame" is a soy ingredient that is contained in this food.

FALCPA did not require precautionary advisory labeling such as "may contain milk" or "made in a facility that also processes peanuts." However, it likely played a role in the proliferation of such advisory labeling on packaged foods that has occurred since then. Such precautionary labeling is used voluntarily by industry and is not currently regulated in the U.S. (Allen et al. 2014), but the lack of consistency in wording and usage has led to considerable confusion and criticism (Zurzolo et al. 2012). However, the presence of the precautionary labeling, in particular, may have had an upside: its near ubiquity has likely raised general awareness of the seriousness of food allergies and highlighted the difficulty of eliminating food allergen cross-contact during product manufacturing. In addition, the presence of precautionary labeling may protect the food-allergic customer from allergens present due to cross-contact because FALCPA only covers allergens present as intended ingredients. FDA has advised that any advisory statements such as "may contain [allergen]" must be truthful and not misleading, and advisory statements should not be a substitute for adherence to current Good Manufacturing Practices (FDA 2006).

3.2.3   U.S. FDA Food Code as a Foundation to State Food Code Laws

The U.S. FDA Food Code is a model law published by the FDA that recommends detailed food safety procedures for food service establishments, restaurants, retail food stores, nursing homes, and child care facilities. The FDA Food Code is not itself a law, but it serves as a reference that state and local governments can adopt as their own without the need for each to create their own from scratch (FDA 2005). The Food Code is now updated every four years to reflect the latest food safety knowledge, including newly identified or prioritized risks.

Food allergens were not explicitly addressed in the Food Code until 2005. Table 3.1 illustrates when provisions related to food allergens were added to the Food Code. Reflecting the 2004 passage of FALPCA, the 2005 version included definitions of the major food allergens and added labeling requirements for packaged foods. Chapter 2 Subpart 102 of the 2005 Food Code also required for the first time that the "person in charge" at a retail food establishment have a demonstrated knowledge of major food allergens and the symptoms of an allergic reaction (FDA 2005). In 2009, the Food Code required the person in charge to include food allergy awareness information in their employee training (FDA 2009). Various training programs have been developed to meet these and other training requirements for food service employees, including the Servsafe® Allergens Online Course

Table 3.1 Food allergen provisions in the FDA food code versions and adoption by states

FDA food code version	Food allergen provisions	States adopting this version(s)
1995–2001	None	AZ, CT, DC, GA[a], ID, IN, LA, MA, MN, NJ, NY, SD, VT
2005	• Added definition of major allergens (referring to the Food Allergen Labeling and Consumer Protection Act) to Chap. 1 • Added a requirement for the person in charge at a food establishment demonstrate knowledge of major food allergens and the symptoms of an allergic reaction to Chap. 2 • Added labeling requirements for major food allergens in packaged foods in Chap. 3 • Updated allergen information in Annex 4 on active managerial control/HACCP principles	AL, AK, CA, GA[a], IL, KY, RI, VA, WV
2009	Added "food allergy awareness" to Chap. 2 as a part of the food safety training of employees as related to their duties by the person in charge	AR, CO, FL, HI, IA, KS, ME, MD, MI, MO, NE, NV, NH, NC, ND, OH, OK, OR, TN, UT, WA, WI, WY
2013	Amended Chap. 4 to clarify that food contact surfaces of equipment and utensils that have contacted a raw animal food that is a major food allergen such as raw fish or seafood must be cleaned and sanitized prior to contacting different types of raw animal foods	DE, MS, MT, NM, PA, SC, TX

Sources: AFDO (2016), FDA (2016a, b)
[a]Georgia Department of Agriculture and Georgia Department of Public Health have adopted different versions of the FDA Food Code

(discussed in Chap. 9) and the University of Minnesota Extension's Serve it Up Safely™ Course (discussed in Chap. 12).

All states have adopted some form of the U.S. FDA Food Code, including seven that have adopted the most current version, but 13 states have not adopted a version of the code that incorporates food allergen provisions (Table 3.1). However, some of these states, in particular, Massachusetts, have adopted other laws for restaurants related to food allergens (discussed in Sect. 3.4.1).

3.2.4 Food Safety Modernization Act

The landmark Food Safety Modernization Act (FSMA) of 2011 contributed in several ways to food allergy/allergen awareness and may have influenced state initiatives that occurred around that time.

The Food Allergy and Anaphylaxis Management Act (FAAMA) was originally introduced to the U.S. Congress in 2005 and finally became law as Section 112 of FSMA in 2011. FAAMA required that the U.S. Secretary of Health and Human Services develop and distribute voluntary guidelines for schools to use for food allergy and anaphylaxis management. Through the actions of conscientious lawmakers (including then-Connecticut Senator Chris Dodd, the parent of a child with food allergies, and Maryland Representative Steny Hoyer, who has a grandchild with a food allergy), a bill that seemed destined to languish due to its relatively small breadth was passed by including it in the much broader and higher profile FSMA (Allergic Living 2011).

The guidelines that FAAMA decreed be written were eventually published by the Centers for Disease Control in 2013 as the "Voluntary Guidelines for Managing Food Allergies in Schools and Early Care and Education Programs" (CDC 2013). The guidelines include recommendations on planning, developing, and implementing a comprehensive plan for food allergy management. Priority activities recommended in the guidelines include ensuring daily management for individual children with food allergies, preparation for food allergy-related emergencies, and staff and student training and education. Recommendations for putting the guidelines into place administratively are also included.

These guidelines represent the first federal guidance issued on food allergy management (Allergic Living 2011) and prevent individual school districts from having to start from scratch in developing food allergy management policies. As an additional spur to encourage adoption of food allergy management policies, FAAMA also allowed public school districts that adopt food allergy management guidelines to become eligible for certain incentive grants (although such incentive grants, if they have been awarded, have not been widely publicized).

FSMA will also have a significant impact on the food manufacturing industry. The new Hazard Analysis and Risk-Based Preventive Controls for Human Foods (PCHF) regulations specify food allergens as a chemical hazard that must be assessed and, if necessary, controlled via an appropriate food allergen preventive control (21 CFR Part 117.126). Such controls include procedures to prevent allergen cross-contact and ensure proper labeling of food products. While the PCHF regulations have not yet directly impacted many state laws, it is likely they will in the future. In the meantime, the PCHF regulations will continue to raise awareness of food allergens and the importance that federal agencies place on their control and management.

3.2.5   School Access to Emergency Epinephrine Act of 2013

This federal law was developed to encourage states to pass laws that do the following:

1. Require stock epinephrine in public elementary and secondary schools
2. Allow and train authorized personnel at public elementary and secondary schools to administer epinephrine to someone who is believed to be having an anaphylactic reaction
3. Provide protection against liability ("Good Samaritan" law) against those that administer epinephrine "in good faith" to someone believed to be having an anaphylactic reaction

States are not required to pass such laws, but if they do, they will be given preference in eligibility for asthma education funding; about $22M was awarded towards these grants nationally in 2013 (Dahlman 2013).

3.3   Impetus for State Laws Related to Food Allergy Management

As discussed above, various factors have influenced states to make legislative changes related to food allergy/allergen management, including new federal laws, revisions to the Food Code, and incentives from the federal government. Advocacy efforts of all types (individual, grass roots organizations, nonprofit agencies, and epinephrine injector manufacturers) contributed to states passing laws to improve food allergy management (Johnson and Ho 2016).

Personal experiences with food allergies and anaphylaxis such as Sen. Dodd's and Rep. Hoyer's have become more and more common as food allergies become more prevalent (Branum and Lukacs 2009). Even President Barack Obama's daughter Malia's food allergy was mentioned when Obama signed the School Access to Emergency Epinephrine Act into law (LoGiurato 2013). Unfortunately, in some cases, the awareness of food allergies may stem from tragedies. Such events can, however, spur prompt action, as occurred following Sabrina Shannon's death and in more recent deaths that occurred in U.S. schools (Table 3.2).

Table 3.2 Fatal anaphylaxis reactions prompting U.S. state school epinephrine legislation

Reaction date	Geographic location	Student	Allergen	Response	Response date	References
December 2010	Chicago, Illinois	13-year-old Katelyn Carlson	Peanut allergen in Chinese takeout eaten during a school party	New state law passed to allow schools to stock and administer epinephrine	August 2011	Illinois Government News Network (2011), Rappaport (2012)
January 2012	Chesterfield County, Virginia	7-year-old Ammaria Johnson	Peanut given to her by another student	New state law passed to require schools to implement policies for the possession and administration of epinephrine	April 2012	Brown (2012), Martin (2012)
July 2013	El Dorado County, California	13-year-old Natalie Giorgi	Peanut allergen in a Rice Krispies treat	New state law passed to require all schools to stock at least one epinephrine injector and have one staff member trained in its use	September 2014	Furillo (2014), Caiola (2014)
September 2013	Corpus Christi, Texas	Cameron Espinosa	Fire ants attack during a middle-school football game	Senate Bill 66 (named after Cameron's #66 football jersey) passed into law to allow all Texas schools to stock epinephrine	May 2015	Lira (2015)

In these four examples of fatal anaphylaxis reactions that occurred in school children, rapid changes to state laws were made. The nationwide publicity for some of these tragic events may have prompted other states to take action, as many states enacted legislation related to school epinephrine availability during the time period between 2010 and 2016 (FARE 2016b).

The rest of this chapter reviews state legislation that has been passed related to food allergy/allergen management in restaurants, schools, other public entities, on emergency vehicles, and also in insurance coverage requirements for hypoallergenic formula for infants and children with food allergies.

3.4 Areas of State Legislation Related to Food Allergy/Allergen Management

3.4.1 Food Allergen Management in Restaurants

Beyond food allergen-related updates to state Food Codes (discussed in Sect. 3.2.3 above), at least five states (Massachusetts, Maryland, Michigan, Rhode Island, and Virginia) have passed other laws specifically designed to improve food allergen management in restaurants. Massachusetts led the way in 2009 when it passed The Act Relative to Food Allergy Awareness in Restaurants, followed by similar requirements in Rhode Island in 2012. Table 3.3 summarizes state food allergy related laws that have been passed through July 2016 (FARE).

Overall, the states have similar provisions, with some of the common themes being food allergen training for management or posting signage for employees to avoid cross contact or for customers to alert staff to their food allergies. One state (Rhode Island) passed a law to create a registry of allergen-friendly restaurants. Creating such registry might encourage more restaurants to strive to be part of the

Table 3.3 State requirements for restaurant food allergy training and awareness

State	Year passed	Reference	Restaurant food allergy training and awareness requirements
Maryland	2015	Maryland General Assembly Department of Legislative Services (2015)	• Restaurants must display a food allergy awareness poster in the staff area
Massachusetts	2009	Foley (2011)	• Restaurants must display a food allergy awareness poster in the staff area and include a menu notice for customers to inform servers of food allergies • Restaurants must have a food protection manager who has completed allergen awareness training
Michigan	2015	Michigan Legislature (2014)	• Food safety managers must take a training course with an allergen awareness component
Rhode Island	2012	Rhode Island General Assembly (2012)	• Food allergy awareness posters must be displayed in staff areas, and menus must remind customers to notify server of food allergies • Restaurant managers must be trained on food allergies
Virginia	2015	Virginia Legislature (2015)	• Food allergy awareness is required in restaurant training standards • Training materials on food allergy awareness must be provided to restaurants

listing, but may be difficult to establish, as evidenced by Rhode Island's continuing delay in publishing their registry.

These initiatives are still new, and improvements may be needed. Some advocates have expressed concern that requirements for customers to notify restaurant staff of food allergies may elicit a false sense of security for the customers, although data is not yet available to tell whether this will be a significant concern or not (AllergyEats 2013). In addition, arguments have been made that more extensive food allergen management training should be required, and some training should be required for all restaurant staff, not just managers (AllergyEats 2010).

Restaurants are also considered public entities, and many states have enacted legislation allowing public entities to maintain undesignated (stock) epinephrine, as discussed in Sect. 3.4.3.

3.4.2   Food Allergy Management in Schools

In 1998, two students in Omaha, Nebraska, died from asthma attacks while attending public school (Barclay 2006; Murphy et al. 2006). These events galvanized the community to develop an emergency response protocol for schools to use to prevent such tragedies from occurring (Murphy et al. 2006). In addition to addressing asthma attacks, the protocol was broadened to cover responses to anaphylaxis and included stocking epinephrine in schools and administration of epinephrine by nurses or designated trained nonmedical staff as part of the protocol. Eventually, the emergency response protocol was adopted by the Nebraska State Board of Education in 2003 and became mandated by state regulation to be used in all Nebraska schools in 2003 (Murphy et al. 2006; Murphy 2006).

The Commonwealth of Massachusetts followed a similar path at about the same time, issuing regulations dealing with training and use of epinephrine in schools and publishing detailed guidelines for management of food allergies in schools in 2002 (Sheetz et al. 2004).

Building on these and other early efforts, tremendous progress has been made in recent years towards improving food allergy management in schools, largely through passage of state laws that involve (1) developing food allergy management policies for schools or plans for individual students or (2) improving access to epinephrine.

Most (~75%) states now have laws that require school districts to develop policies to track students with severe allergy and to ensure that their health records are periodically updated. Slightly fewer states (~69%) require schools to have policies in place for handling anaphylaxis events, while less than half of the states require schools to document anaphylaxis or epinephrine administration events (AAFA 2016). Sixteen states currently have published statewide guidelines for school food allergy management (FARE 2017).

Rapid progress has been made in improving epinephrine access in U.S. schools. All states and the District of Columbia have now passed legislation allowing students to carry and self-administer epinephrine, with laws differing in whether they require a documented prescription, physician authorization, training of documentation, and parental permission and release of school from liability for students to carry and self-administer epinephrine (The Network for Public Health Law 2013). However, not all students know they have a food allergy and therefore won't have an epinephrine prescription, as evidenced by the large percentage (25–58% in various studies) of anaphylactic reactions occurring in schools occur in individuals with no prior history of allergies (McIntyre et al. 2005; Illinois State Board of Education 2016). Many states have therefore moved to allow undesignated epinephrine to be stocked in schools. Currently, all states except Hawaii permit stock epinephrine to be available within schools (AAFA 2016), with 16 schools passing such legislation in 2013 alone (CBS News 2013). Eleven states (Arizona, California, Connecticut, Delaware, Maryland, Michigan, Nebraska, Nevada, New Jersey, North Carolina, and Virginia) and the District of Columbia have passed laws that go a step further: they explicitly require schools to stock epinephrine (FARE 2016b).

States differ in whether they require a school nurse or whether a trained staff member can perform the administration of epinephrine. Laws releasing liability for those administering asthma and anaphylaxis medications to individuals at schools are in place in many states (AAFA 2016; Baulsir and Inniss 2015).

Additional school food allergy management measures have also been considered, including food allergen avoidance polices in Michigan (Portnoy and Shroba 2014) and a swiftly repealed 2007 Rhode Island law prohibiting the sale of peanuts or tree nuts in school cafeterias (Coleman 2008). Other recommendations proposed to improve food allergy management in schools include increasing the number of school nurses and providing funding for schools to implement policies.

3.4.3   Food Allergy Management in Other Public Locations/Public Entities

In addition to schools, the potential for food allergy reactions to occur in public locations including restaurants, colleges and universities, day care facilities, sports venues, amusement parks, recreation camps, and other similar types of establishments has been recognized. As a result, many states have developed "public entity" laws, analogous to some of the state laws for schools, which allow such establishments to stock and administer undesignated epinephrine.

As of 2016, 27 states have laws that allow entities to stock and administer epinephrine while an additional six states have pending legislation regarding stock epinephrine for entities (McEnrue and Procopio 2016; FARE 2015). States differ on how they define "entity" (although most use a definition that includes organizations where allergens that could trigger anaphylaxis may be present) and who they allow to administer the epinephrine. All of the states have training requirements for those administering the epinephrine, although the training requirements and renewal requirements differ between states (McEnrue and Procopio 2016). Importantly, liability protection is included in most of these laws for the public entity, the trained employee administering the epinephrine in good faith, and the health care professional prescribing and dispensing the stock epinephrine (McEnrue and Procopio 2016).

3.4.4 Food Allergy Management on Emergency Vehicles

States and local laws typically control the equipment and medications that emergency vehicles such as ambulances carry and that emergency personnel are authorized to administer to patients. As discussed in more detail in Chap. 4, only a small number (11) of states require that emergency vehicles carry epinephrine autoinjectors. In addition, many states do not explicitly allow all emergency medical personnel to administer epinephrine or do not require that each emergency vehicle have a crew member authorized to administer epinephrine. These apparent gaps in coverage are issues that the food allergy advocacy organization Food Allergy Research & Education (FARE) is currently seeking to address via state initiatives (FARE 2016a). See Chap. 4 for more details on this topic.

3.4.5 Insurance Coverage for Hypoallergenic Formula

Some infants and children with food allergies cannot consume regular formulas, as the intact proteins within these formulas are allergenic to them. Hypoallergenic formulas can, however, be made from extensively hydrolyzed milk (or soy) proteins, or formulated from the individual amino acids themselves. These products are not allergenic as the epitopes of the allergenic proteins are no longer intact and recognizable by the immune system (Baker et al. 2000).

These hypoallergenic formulas are significantly more expensive than regular formulas (Baker et al. 2000). Due to their expense, insurance coverage for these formulas, which are generally classified as medical foods, is available in some states (APED 2017).

Currently 19 states require insurance coverage for medical foods (including hypoallergenic infant formula that is prescribed by a doctor), including Arizona, Connecticut, Illinois, Kentucky, Maine, Maryland, Massachusetts, Minnesota, Missouri, Nebraska, New Hampshire, New Jersey, New York, Oregon, Pennsylvania, Rhode Island, South Dakota, Texas, and Washington (APED 2017). In several states, Special Supplemental Nutrition Program for Women, Infants, and Children (WIC) or Medicaid will also help pay for these special formulas if required. Different states have different Medicaid and WIC coverage, and navigating the reimbursement options for hypoallergenic infant formulas may require working directly with the companies that manufacture and market the products (who may supply the formula at no cost if no other reimbursement is available) (Neocate 2017).

3.5 Summary and Conclusions

In the last few decades, significant progress has been made towards improving food allergy/allergen management in schools, restaurants, and elsewhere across the U.S., with much of this progress attributable to initiatives and legislation arising at the state level. These state efforts have enhanced food allergy awareness, provided guidelines for food allergen management, and improved access to medical products (especially autoinjectable epinephrine, but also hypoallergenic formula) that make life safer for those with food allergies.

Federal efforts in recent years built a foundation for these state initiatives by providing civil rights protections for those with food allergies and creating model guidelines for states and school districts to draw from when developing food allergen management food codes and food allergy management policies. In some cases, the federal government provided financial incentives for states and school districts to stock epinephrine in more schools or put food allergen management plans in place. Other federal efforts such as the FALCPA and FSMA did not directly impact state efforts, but improved consumer safety and increased general awareness of food allergens. FSMA has helped food allergens receive the same level of attention previously reserved for foodborne pathogens.

State legislation related to food allergy management can be classified into two categories: laws seeking to prevent allergic reactions and those attempting to improve responses to food allergy emergencies. An increasing number of states have begun passing legislation mandating food allergen training for restaurant managers and/or requiring signage to remind restaurant staff of the hazards of cross-contact or remind customers to mention their food allergies.

To improve responses to food allergy emergencies, growing numbers of states have developed statewide guidelines for food allergy management in schools, while many states sponsor or provide funding for staff training in food allergies (AAFA 2016). All states now allow students to carry and self-administer epinephrine, and as of this writing all states except Hawaii permit schools to stock epinephrine. Eleven states go further, requiring schools to stock epinephrine (FARE 2016b). Following the success that occurred in schools, states are increasingly allowing restaurants and other public entities to maintain stock epinephrine to protect all food-allergic individuals visiting their establishment. Other current state initiatives seek to increase epinephrine availability on emergency vehicles and improve access to hypoallergenic infant formula through insurance coverage or government programs.

Challenges remain. Different laws in different states mean inconsistencies exist, complicating the development of food allergen training that can be used across state lines or determining whether a particular establishment in a multi-state chain may carry stock epinephrine. Awareness of food allergens still needs to be improved, but without building anxiety. Better consistency in advisory labeling may be one small way to do this (Roses 2011), as should careful training of school staff, restaurant employees (and inspectors), and others who may be involved with serving and protecting food allergic individuals.

The costs associated with food allergen training and epinephrine supply are not trivial. The recent, steep price increase for the EpiPen created a cloud over its manufacturer's role in improving epinephrine access in schools (Johnson and Ho 2016), and somewhat controversial attention has been brought to the relatively low frequency of food allergy deaths relative to perception of risk and accompanying anxiety that widespread distribution of epinephrine is thought to trigger (Colver and Hourihane 2006). Much of the progress at all levels (federal, state, and local) has been fueled by efforts of those individuals personally affected by food allergies and organizations such as FARE and Asthma and Allergy Foundation of America, among many others. The passion of food allergy advocates has proven enormously effective, but must be tempered with realism and a focus on finding consensus and identifying end goals in order to achieve successful passage of legislation and concomitant change.

Despite state laws, not all schools have epinephrine available (Shah et al. 2014). Schools in less affluent areas have been shown to be less likely to stock epinephrine (Shah et al. 2014), so more universal requirements for stocking epinephrine or other mechanisms to ensure all children have access epinephrine may be warranted. Similarly, not all students with food allergies have school health management plans, and minority and lower-income students have been shown to be less likely to have a plan than white or higher income students (Gupta et al. 2014). Such inequalities in care for students with food allergies, which have been reviewed previously (McQuaid et al. 2016), must be addressed.

Although it remains early, some data suggests that school allergy management initiatives are successful: stock epinephrine is being administered, often to students who never previously had an anaphylactic reaction (DeSantiago-Cardenas et al. 2015; Virginia Department of Health 2014). School initiatives were among the first legislative efforts attempted, and their success is a promising sign for more nascent initiatives, which have and will continue to draw upon what has been learned from state legislation for food allergy management in schools.

Many of the state and local initiatives related to food allergens were initiated in response to tragic deaths of individuals with food allergies. Improving epinephrine availability may appear to be a quicker fix towards preventing such devastating events. However, food allergen controls to prevent accidental exposure should ultimately decrease the need for emergency responses to allergic reactions. State requirements for food allergen training for food service workers and school staff, food allergen awareness signage in restaurants, and implementation of the current Food Code with the most detailed allergen controls are examples of efforts that should reduce the need for epinephrine and will ultimately save lives. Such measures will also improve the quality of life for those with food allergies and their families by providing more assurance that the food they eat in restaurants, schools, or public places is safe for them.

Acknowledgements The author would like to thank Stephanie Tai, Steven Ingham, Charles Czuprynski, Tong-Jen Fu, and Lauren S. Jackson for reviewing early drafts of this chapter, and Tong-Jen Fu, Lauren S. Jackson, and Kathiravan Krishnamurthy for organizing the conference that led to this chapter.

References

Allen, K.J., P.J. Turner, R. Pawankar, S. Taylor, S. Sicherer, G. Lack, N. Rosario, M. Ebisawa, G. Wong, E.N.C. Mills, K. Beyer, A. Fiocchi, and H.A. Sampson. 2014. Precautionary labelling of foods for allergen content: Are we ready for a global framework? *The World Allergy Organization Journal.* 7: 10.

Allergic living. 2011. FAAMA: Inside the U.S. school allergy law. Available at: http://allergicliving.com/2011/01/12/qa-faama-school-allergy-law/. Accessed 20 January 2017.

AllergyEats. 2010. Massachusetts food allergy awareness law goes into effect... But is it enough? Available at: https://www.allergyeats.com/massachusetts-food-allergy-law/. Accessed 27 February 2017.

———. 2013. State food allergy laws must progress, not stand still. Available at: https://www.allergyeats.com/state-food-allergy-laws-must-progress-not-stand-still/. Accessed 27 February 2017.

APED (American Partnership for Eosinophilic Disorders). 2017. State insurance mandages for elemental formula. Available at: http://apfed.org/advocacy/state-insurance-mandates-for-elemental-formula/. Accessed 20 January 2016.

AFDO (Association of Food and Drug Officials). 2016. Real progress in Food Code adoption. Available at: http://www.fda.gov/downloads/Food/GuidanceRegulation/RetailFoodProtection/FoodCode/UCM476819.pdf. Accessed 20 January 2017.

AAFA (Asthma and Allergy Foundation of America). 2016. 2016 State honor roll: Asthma and allergy policies for schools. Available at: http://www.aafa.org/media/2016-State-Honor-Roll-Report-Asthma-Allergy-Policies-in-Schools.pdf. Accessed 18 January 2017.

Baker, S.S., W.J. Cochran, F.R. Greer, M.B. Heyman, M.S. Jacobson, T. Jaksic, N.F. Krebs, A.E. Smith, D.E. Yuen, W. Dietz, E. Yetley, S.S. Harris, A. Prendergast, G. Grave, V.S. Hubbard, D. Blum-Kemelor, R.M. Lauer, S.C. Denne, R. Kleinman, P. Kanda, and N. Comm. 2000. Hypoallergenic infant formulas. *Pediatrics.* 106: 346–349.

Barclay, L. 2006. School emergency response program for asthma: A newsmaker interview with Kevin R. Murphy, MD. Available at: http://www.medscape.com/viewarticle/528479. Accessed 19 January 2016.

Baulsir, B., and B. Inniss. 2015. Summary matrix of state laws addressing Epi-Pen use in schools. Available at: https://www.networkforphl.org/_asset/wh2271/Food-Allergy-Epi-Pen-50-State-Compilation-FINAL.pdf. Accessed 18 January 2017.

Branum, A.M., and S.L. Lukacs. 2009. Food allergy among children in the United States. *Pediatrics.* 124: 1549–1555.

Brown, E. 2012. Virginia first-grader Ammaria Johnson dies after allergic reaction. *The Washington Post.* 5 January 2012. Available at: https://www.washingtonpost.com/blogs/virginia-schools-insider/post/virginia-first-grader-ammaria-johnson-dies-after-allergic-reaction/2012/01/05/gIQAefDRdP_blog.html?utm_term=.1d2bd18e9d51. Accessed 4 June 2017.

Caiola, S. 2014. New law requires California schools to stock epinephrine injectors for allergic children. *The Sacramento Bee.* 12 November 2014. Available at: http://www.sacbee.com/news/local/health-and-medicine/healthy-choices/article3873646.html. Accessed 4 June 2017.

CBS News. 2013. More states pass laws to store EpiPens at schools. Available at: http://www.cbsnews.com/news/more-states-pass-laws-to-store-epipens-at-schools/. Accessed 27 February 2017.

CDC (Centers for Disease Control and Prevention). 2013. *Voluntary guidelines for managing food allergies in schools and early care and education programs.* Washington, DC: U.S. Department of Health and Human Services.

Cicutto, L., B. Julien, N.Y. Li, N.U. Nguyen-Luu, J. Butler, A. Clarke, S.J. Elliott, L. Harada, S. McGhan, D. Stark, T.K. Vander Leek, and S. Waserman. 2012. Comparing school environments with and without legislation for the prevention and management of anaphylaxis. *Allergy.* 67: 131–137.

Coleman, S. 2008. Peanut/Tree nut bans in schools. Available at: https://www.cga.ct.gov/2008/rpt/2008-R-0472.htm. Accessed 27 February 2017.

Collins, S.C. 2014. Food allergy management in restaurants: More resources available to keep customers safe. *Today's Dietitian.* 16: 18.

Colver, A., and B. Hourihane. 2006. For and against - Are the dangers of childhood food allergy exaggerated? *British Medical Journal.* 333 (7566): 494–498A.

Dahlman, G. 2013. The promise of the School Access to Emergency Epinephrine Act. FARE Blog. Available at: https://blog.foodallergy.org/2013/12/11/the-promise-of-the-school-access-to-emergency-epinephrine-act/. Accessed 4 June 2017.

DeSantiago-Cardenas, L., V. Rivkina, S.A. Whyte, B.C. Harvey-Gintoft, B.J. Bunning, and R.S. Gupta. 2015. Emergency epinephrine use for food allergy reactions in Chicago Public Schools. *American Journal of Preventive Medicine.* 48: 170–173.

Eigenmann, P.A., and S.A. Zamora. 2002. An internet-based survey on the circumstances of food-induced reactions following the diagnosis of IgE-mediated food allergy. *Allergy.* 57 (5): 449–453.

Fleming, J.T., S. Clark, C.A. Camargo, and S.A. Rudders. 2015. Early treatment of food-induced anaphylaxis with epinephrine is associated with a lower risk of hospitalization. *Journal of Allergy and Clinical Immunology: In Practice.* 3: 57–62.

Foley, K. 2011. Letter to local boards of health re: enforcement guidelines for allergen awareness regulation. Available at: http://www.mass.gov/eohhs/docs/dph/environmental/foodsafety/food-allergen-1-enforcement-guidelines.pdf. Accessed 27 February 2017.

FARE (Food Allergy Research & Education). 2017. School guidelines. Available at: https://www.foodallergy.org/laws-and-regulations/guidelines-for-schools. Accessed 27 February 2017.

Food Allergy Canada. 2017. Sabrina's law. Available at: http://foodallergycanada.ca/resources/sabrinas-law/. Accessed 20 January 2017.

FDA (U.S. Food and Drug Administration). 2004. Food Allergen Labeling and Consumer Protection Act of 2004 (Public Law 108-282, Title II). Available at: https://www.fda.gov/food/guidanceregulation/guidancedocumentsregulatoryinformation/allergens/ucm106187.htm. Accessed 4 June 2017.

———. 2005. Food Code 2005. Available at: http://www.fda.gov/Food/GuidanceRegulation/RetailFoodProtection/FoodCode/ucm2016793.htm. Accessed 25 March 2015.

———. 2006. Guidance for industry: Questions and answers regarding food allergens, including the Food Allergen Labeling and Consumer Protections Act of 2004. Available at: http://www.fda.gov/food/guidanceregulation/guidancedocumentsregulatoryinformation/allergens/ucm059116.htm. Accessed 7 February 2017.

———. 2009. Food Code 2009. Available at: http://www.fda.gov/Food/GuidanceRegulation/RetailFoodProtection/FoodCode/ucm2019396.htm. Accessed 17 June 2016.

———. 2016a. FDA Food Code. Available at: http://www.fda.gov/Food/GuidanceRegulation/RetailFoodProtection/FoodCode/default.htm. Accessed 20 January 2017.

———. 2016b. State retail and food service codes and regulations by state. Available at: http://www.fda.gov/Food/GuidanceRegulation/RetailFoodProtection/FoodCode/ucm122814.htm. Accessed 20 January 2017.

FARE (Food Allergy Research & Education). 2015. Public access to epinephrine. Available at: http://www.foodallergy.org/advocacy/advocacy-priorities/access-to-epinephrine/public-access-to-epinephrine. Accessed 22 March 2016.

———. 2016a. EMTs and epinephrine. Available at: http://www.foodallergy.org/advocacy/ems. Accessed 20 January 2017.

————. 2016b. School access to epinephrine map. Available at: http://www.foodallergy.org/advocacy/epinephrine/map. Accessed 5 December 2016.

Furillo, A. 2014. Family sues city after girl's peanut-allergy death at Camp Sacramento. *The Sacramento Bee*. April 18, 2014 edition. Available at: http://www.sacbee.com/news/local/health-and-medicine/article2596198.html. Accessed 4 June 2017.

Gregory, N.L. 2012. The case for stock epinephrine in schools. *NASN School Nurse*. 27: 223–225.

Gupta, R.S., J.S. Kim, E.E. Springston, B. Smith, J.A. Pongracic, X.B. Wang, and J. Holl. 2009. Food allergy knowledge, attitudes, and beliefs in the United States. *Annals of Allergy Asthma and Immunology*. 103: 43–50.

Gupta, R.S., V. Rivkina, L. DeSantiago-Cardenas, B. Smith, B. Harvey-Gintoft, and S.A. Whyte. 2014. Asthma and food allergy management in Chicago Public Schools. *Pediatrics*. 134 (4): 729–736.

Illinois Government News Network. 2011. Governor Quinn signs bill to expand emergency access to life-saving allergy medicine for children. Available at: http://www3.illinois.gov/PressReleases/ShowPressRelease.cfm?SubjectID=3&RecNum=9640. Accessed 27 February 2017.

Illinois State Board of Education. 2016. Report of use of undesignated epinephrine school year 2014-15. Available at: http://www.isbe.net/pdf/school_health/epinephrine-use-report14-15.pdf. Accessed 5 December 2016.

Johnson, C. Y., and C. Ho. 2016. How Mylan, the maker of EpiPen, became a virtual monopoly. *The Washington Post*. August 25, 2016 edition. Availabe at: https://www.washingtonpost.com/business/economy/2016/08/25/7f83728a-6aee-11e6-ba32-5a4bf5aad4fa_story.html?utm_term=.4f53427a8b70. Accessed 4 June 2017.

Lira, J. 2015. Cameron Espinosa Act expected to be signed into law. Available at: http://www.kristv.com/story/29070420/cameron-espinosa-act-expected-to-be-signed-into-law. Accessed 19 July 2016.

LoGiurato, B. 2013. If you have a peanut allergy, Obama just signed a bill that will make your life a whole lot easier. *Business Insider*. November 15, 2013 edition. Available at: http://www.businessinsider.com/peanut-allergies-allergy-epi-pen-bill-malia-obama-sign-2013-11. Accessed 4 June 2017.

Lynch, W. 2017. The rights of individuals with allergy-related disabilities under the ADA. Available at: http://www.foodallergy.org/file/ada-webinar-slides.pdf. Accessed 22 March 2017.

Manivannan, V., R.J. Hyde, D.G. Hankins, M.F. Bellolio, M.G. Fedko, W.W. Decker, and R.L. Campbell. 2014. Epinephrine use and outcomes in anaphylaxis patients transported by emergency medical services. *American Journal of Emergency Medicine*. 32: 1097–1102.

Martin, C. 2012. Virginia enacts stock epinephrine law to help protect students with allergies. *Forbes*. 27 April 2012 edition. Available at: https://www.forbes.com/sites/work-in-progress/2012/04/27/virginia-enacts-stock-epinephrine-law-to-help-protect-students-with-allergies/#738a6fea4259. Accessed 4 June 2017.

Maryland General Assembly Department of Legislative Services. 2015. House Bill 751: Health - Food allergy awareness. Available at: mgaleg.maryland.gov/2015RS/fnotes/bil_0001/hb0751.pdf. Accessed 22 June 2017.

McEnrue, M., and V. Procopio. 2016. Epinephrine stocking laws in the U.S. Available at: https://www.networkforphl.org/_asset/8483ms/Issue-Brief-Epi-Entity-Stocking.pdf. Accessed 18 January 2017.

McIntyre, C.L., A.H. Sheetz, C.R. Carroll, and M.C. Young. 2005. Administration of epinephrine for life-threatening allergic reactions in school settings. *Pediatrics*. 116: 1134–1140.

McQuaid, E.L., M.L. Farrow, C.A. Esteban, B.N. Jandasek, and S.A. Rudders. 2016. Topical review: Pediatric food allergies among diverse children. *Journal of Pediatric Psychology*. 41: 391–396.

Michigan Legislature. 2014. Public Act 516 of 2014. Available at: http://legislature.mi.gov/doc.aspx?2013-SB-0730. Accessed 22 June 2017.

Murphy, K.R. 2006. Administration of epinephrine for life-threatening allergic reactions in school settings. *Pediatrics*. 117: 1862–1862.

Murphy, K.R., R.J. Hopp, E.B. Kittelson, G. Hansen, M.L. Windle, and J.N. Walburn. 2006. Life-threatening asthma and anaphylaxis in schools: A treatment model for school-based programs. *Annals of Allergy Asthma & Immunology*. 96: 398–405.

Neocate, N. 2017. Reimbursement. Available at: http://www.neocate.com/reimbursement/. Accessed 20 January 2016.

O'Brien-Heinzen, T. 2010. A complex recipe: Food allergies and the law. *Wisconsin Lawyer*. 83 (5).

Portnoy, J.M., and J. Shroba. 2014. Managing food allergies in schools. *Current Allergy and Asthma Reports*. 14: 7.

Rappaport, L. 2012. Finding food allergy allies: Schools, states, restaurants take steps; beyond the peanut-free table. Available at: http://www.wsj.com/articles/SB10001424052970203918304577243554276460014. Accessed 27 February 2017.

Rhode Island General Assembly. 2012. An act relating to health and safety - Food allergy awareness in restaurants. Available at: http://webserver.rilin.state.ri.us/PublicLaws/law12/law12414.htm. Accessed 22 June 2017.

Roses, J.B. 2011. Food allergen law and the Food Allergen Labeling and Consumer Protection Act of 2004: Falling short of true protection for food allergy sufferers. *Food Drug Law Journal*. 66: 225–242.

Shah, S.S., C.L. Parker, E.O. Smith, and C.M. Davis. 2014. Disparity in the availability of injectable epinephrine in a large, diverse U.S. school district. *Journal of Allergy and Clinical Immunology: In Practice* 2: 288–293.e1.

Sheetz, A.H., P.G. Goldman, K. Millett, J.C. Franks, C.L. McIntyre, C.R. Carroll, D. Gorak, C.S. Harrison, and M. Abu Carrick. 2004. Guidelines for managing life-threatening food allergies in Massachusetts schools. *Journal of School Health*. 74: 155–160.

Terry, J. 2014. New laws improve access to epinephrine in NYS schools. Available at: http://www.allergyadvocacyassociation.org/index.php/87-advocacy-stories/230-new-laws-improve-access-to-epinephrine-in-nys-schools-advocacy. Accessed 20 January 2017.

The Network for Public Health Law. 2013. Food safety - Food allergy policy issue brief. Available at: https://www.networkforphl.org/_asset/mtb1lf/Food-Allergy-Policy-Project-Issue-Brief.pdf. Accessed 27 February 2017.

Virginia Department of Health. 2014. Epinephrine report. Available at: http://www.vdh.virginia.gov/. Accessed 22 June 2017.

Virginia Legislature. 2015. HB 2090 Restaurants; training standards that address food safety and food allergy awareness and safety, Available at: https://lis.virginia.gov/cgi-bin/legp604.exe?151+sum+HB2090. Accessed 4 June 2017.

Weiss, C., and A. Munoz-Furlong. 2008. Fatal food allergy reactions in restaurants and food-service establishments: Strategies for prevention. *Food Protection Trends*. 28: 657–661.

Xu, Y.S., M. Kastner, L. Harada, A. Xu, J. Salter, and S. Waserman. 2014. Anaphylaxis-related deaths in Ontario: A retrospective review of cases from 1986 to 2011. *Allergy, Asthma and Clinical Immunology*. 10: 8.

Yee Ming, L., and X. Hui. 2015. Food allergy knowledge, attitudes, and preparedness among restaurant managerial staff. *Journal of Foodservice Business Research*. 18 (5): 454–469.

Zurzolo, G.A., M.L. Mathai, J.J. Koplin, and K.J. Allen. 2012. Hidden allergens in foods and implications for labelling and clinical care of food allergic patients. *Current Allergy and Asthma Reports*. 12: 292–296.

Chapter 4
Current Trends in Food Allergy Advocacy: Prevention, Preparedness, and Epinephrine Availability

Jennifer Jobrack

4.1 Introduction

Food allergies are a significant public health issue, affecting between 2 and 10% of the general population (Chafen et al. 2010). This prevalence appears to be increasing, with the CDC estimating that food allergies in children increased 18% within a recent 10-year period (Branum and Lukacs 2009). Food allergies contributed to 11,000 hospitalizations in addition to 164,000 emergency department visits in 2006 and 2007 (Patel et al. 2011). The annual economic burden of food allergies in the U.S. has been estimated to be almost $25 billion (Gupta et al. 2013). Beyond the statistics are the effects that food allergies have on the quality of life for those with food allergies and their families, which have recently been reviewed (Walkner et al. 2015). The need for constant vigilance during many everyday activities causes stress, especially for children who may be excluded from activities or actively bullied because of their food allergy.

Exposure to an allergen can very rapidly elicit a variety of reactions in an allergic individual. Anaphylaxis, the most serious of these reactions, is a systemic response to the allergen that can be life threatening. Anaphylaxis may affect many organs, resulting in dangerously reduced blood pressure, cardiovascular collapse, and swelling of the tongue and respiratory tissues that can hamper breathing. Epinephrine is the first-line treatment for anaphylaxis as it blocks the effects of the biochemical mediators responsible for the anaphylactic reaction (Song and Lieberman 2015). Treatment with intramuscular epinephrine and avoiding allergen exposure are key strategies used to manage food allergy reactions.

This chapter discusses recent efforts addressing some of the most serious concerns facing those with food allergies and their families. The role of advocacy

J. Jobrack (✉)
Food Allergy Research & Education (FARE),
8707 Skokie Boulevard, Suite 104, Skokie, IL 60077, USA
e-mail: jjobrack@foodallergy.org

© Springer International Publishing AG 2018
T.-J. Fu et al. (eds.), *Food Allergens*, Food Microbiology and Food Safety,
DOI 10.1007/978-3-319-66586-3_4

organizations such as Food Allergy Research & Education (FARE) in both promoting legislative approaches and increasing public awareness is highlighted, and remaining challenges are described.

4.2 Challenges for Food Allergic Individuals and Their Families

Among the biggest challenges currently for people with food allergies is simply understanding the complexity and potential severity of food allergies (Gupta et al. 2008). Initially upon diagnosis, individuals with food allergies may not understand what it means to have a food allergy. They may have had a mild reaction to a food the first time, which mistakenly leads them to believe they have a "mild food allergy". Unfortunately, there is no such thing as a "mild food allergy"; any allergy, even one that has only manifested itself in mild reactions previously, could trigger a serious reaction. Even when carrying an epinephrine autoinjector (EAI), food allergic individuals experiencing a reaction may hesitate to use them. They may believe that the reaction is not serious enough to warrant the use of epinephrine, or they may feel uncomfortable delivering the injection (Song and Lieberman 2015; Dudley et al. 2015).

Reading food product labels is another challenge for food allergic patients and their families (Zurzolo et al. 2016). Knowing what is in the food you or your child is consuming is vital to preventing exposure to food allergens. Advisory labeling is voluntary, and its absence does not mean a food is allergen free (Zurzolo et al. 2016). Advisory labeling is not standardized and may also be less than clear. For example, what does "may contain" mean in terms of the likelihood of allergen presence in that food (Pieretti et al. 2009)? The ambiguity and overuse of advisory labeling (sometimes unfortunately in lieu of having proper manufacturing controls in place to prevent contamination) may have fostered cavalier attitudes among consumers towards these warnings (Hefle et al. 2007).

Another challenge facing food allergic individuals, especially children, is social isolation (Walkner et al. 2015; Portnoy and Shroba 2014). Food allergies are different from other illnesses in that the behavior of other people can directly affect the appearance of symptoms. Schools, daycares, and other organizations must ask other families to cooperate at times, which can create misunderstanding, confusion, and resentment. Children with food allergies may be inadvertently singled out or excluded from activities. Parents of children with food allergies often experience anxiety that can be transferred to their children (Akeson et al. 2007).

By increasing awareness of food allergies and ensuring life-saving epinephrine is available in locations where children may spend time without parental supervision, some of these concerns and challenges may be lessened. Advocates for those with food allergies have been instrumental in increasing awareness of the condition and in finding ways through legislation to improve conditions for those with food allergies such as increasing epinephrine availability. Food allergy advocacy organizations have been instrumental both in the U.S. and in other countries in facilitating governmental rule-making that enhances the safety and quality of life of those with food allergies.

FARE is the leading advocacy organization dedicated to food allergy in the U.S. It was formed in 2012 through the merger of the Food Allergy & Anaphylaxis Network (FAAN) and the Food Allergy Initiative (FAI). The mission of FARE is to "to improve the quality of *life* and the *health* of individuals with food allergies and to provide them *hope* through the promise of new treatments" (FARE 2016a). FARE plays an active role in food-allergy related issues such as school policies, food labeling, restaurant regulations, emergency services, and transportation.

Current legislative initiatives that FARE supports are described in the remaining sections of this chapter, including prevention, preparedness, and epinephrine availability in schools, public places, restaurants, colleges and universities, and on emergency vehicles and airlines.

4.3   Food Allergy Management in U.S. Schools

FDA's 2011 Food Safety Modernization Act (FSMA)'s emphasis on preventive approaches to food safety extends to food allergy planning in schools as well as many other areas of food safety. A component of FSMA is the Food Allergy and Anaphylaxis Management Act (FAAMA). FAAMA was spearheaded by FARE's predecessor, FAAN, and became law as part of FSMA in January 2011. FAAMA required the federal government to create voluntary national guidance material to manage food allergies in schools and related organizations. The Centers for Disease Control and Prevention (CDC) worked with the U.S. Department of Education in collaboration with FARE and others to develop these guidelines, which are available online (CDC 2013).

The CDC guidelines are intended to guide schools into developing food allergy policies of their own. Such policies may include education and training for school staff detailed procedures for handling food allergy reactions, guidelines to prevent student exposure to allergens, and care plans for individual students with allergies. Having a formal food allergy policy in schools can help depersonalize the issue for students with allergies; the family does not have to educate the school on how to protect their child, and the child may be less likely to be singled out as being different. Developing a school food allergy policy requires significant effort. A number of states have implemented statewide guidelines for food allergy management in schools to simplify and unify school food allergy management plans; links to these guidelines are available on FARE's website (FARE 2016e).

4.3.1   The Problem

Food allergies are a tremendous problem in U.S. schools, directly affecting about 1 in 13 children, or "2 in every classroom" (Gupta et al. 2011). Although data have not always been systematically collected over a period of years, there is indication that food allergy prevalence is increasing (Branum and Lukacs 2008).

In addition to increasing prevalence, other challenges exist to food allergy management in schools. School teaching and administrative staff may have tremendous knowledge gaps related to food allergies. On a basic level, they may not know which students have allergies, what an anaphylactic reaction in response to a food allergen looks like, or how to treat anaphylaxis. They may not understand how having a food allergy affects a student's quality of life or realize that biphasic reactions (where a secondary reaction occurs minutes to hours after an initial reaction has subsided) may occur (Agarwal and Yu 2015).

Complicating matters for schools, a significant proportion of children does not know they have a food allergy before they start school, and may have their first allergic reaction during school. A Massachusetts study concluded that 24% of students who developed anaphylactic reactions in schools were unaware that they had a food allergy (McIntyre et al. 2005). Another report found that a similar percentage (25%) of children with peanut or tree nut allergies experienced their first reaction in school (Sicherer et al. 2001).

A key part of a food allergy policy in many schools is ensuring availability of treatment for anaphylaxis. Epinephrine is "safe" and standard medical practice (Sheikh et al. 2009) for anaphylaxis (Kemp et al. 2008) and is the first-line therapy for anaphylaxis treatment, even in children (Farbman and Michelson 2016). A recent medical journal editorial states that there is no absolute contraindication to epinephrine administration to treat anaphylaxis (Dudley et al. 2015). Delays in epinephrine administration increase the risk of a fatal reaction (Sampson et al. 1992), while giving other drugs such as antihistamines or glucocorticoids instead of epinephrine is not recommended (although their use together with epinephrine has been advocated) (Farbman and Michelson 2016).

Among children with food allergies, 18% experienced a reaction at school during the past 2 years (Nowak-Wegrzyn et al. 2001). Even if students were previously diagnosed with a food allergy and have a personal prescription for an EAI, various barriers may prevent their use at school. However, in recent years this has changed, and now students in all states, with appropriate consent, may carry a prescribed EAI (FARE 2016c).

It is recommended that people diagnosed with food allergy carry two EAIs with them at all times. Even if permitted to carry an EAI, school-age children may forget or not want to carry their own EAI. One study found that 25% of children 5 years of age and older did not have their EAI with them at lunch (DeMuth and Fitzpatrick 2011). Nearly half (49%) of reported anaphylactic events that happened in schools occurred in high school students in another study (White et al. 2016). Adolescents are at a higher risk for a life-threatening allergic event, in part because they don't always carry self-injectable epinephrine (Spina et al. 2012; Sampson et al. 2006).

Even if they do have an EAI, it might have expired (Spina et al. 2012) or they may require an additional dose. Increasing costs for EAI may also make it difficult for parents to afford to purchase (and keep purchasing as they expire) EAIs, especially if their child has only had mild reactions previously.

Stock (also called undesignated, unassigned, or general use) EAI that is available at schools for general use whenever anyone has an anaphylactic reaction would clearly solve many of these problems. However, various hurdles have existed that

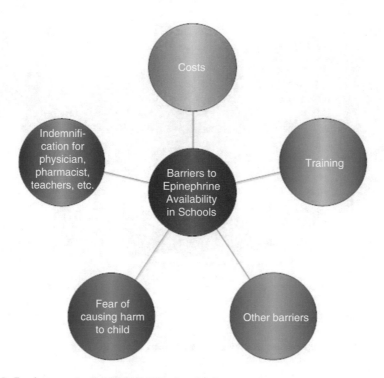

Fig. 4.1 Barriers to epinephrine availability in schools

have made it difficult for stock EAI to be available in schools (Odhav et al. 2015). Obstacles include concerns related to indemnification (Odhav et al. 2015), epinephrine costs, the need for and availability of training for those giving the injections (Chokshi et al. 2015), fear of causing harm to children, and other barriers (Fig. 4.1).

4.3.2 Ensuring Availability of Stock Epinephrine in Schools

FARE has played an important role in ensuring the availability of stock epinephrine in schools throughout the U.S. by facilitating and promoting legislation that removed barriers to allowing epinephrine in schools. FARE has also supported legislative efforts towards requiring schools to stock epinephrine.

These efforts have been successful. Prior to 2010, only five states allowed schools to stock epinephrine in schools. As of July 2016, all states except Hawaii have guidelines or statutes that specifically allow stock epinephrine in schools.

Following on FAAMA, the 2013 School Access to Emergency Epinephrine Act provides preferential consideration in receiving asthma-related grants for child health services to those states that require stock epinephrine to be available at schools (Dahlman 2013). As of July 2016, 11 states now specifically require stock epinephrine in schools (dark green states in Fig. 4.2).

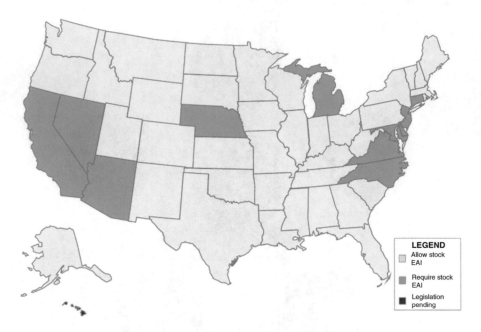

Fig. 4.2 School epinephrine availability by state (adapted from FARE)

Stock epinephrine in schools is being used. Separate from FARE's efforts, the EpiPen4Schools® program has been offered by Mylan Specialty (who markets EpiPen® autoinjectors) to improve availability of epinephrine in schools (Mylan Specialty L.P. 2016). The EpiPen4Schools survey of schools collected data on stock epinephrine use during the 2013–2014 school year (White et al. 2015). A total of 919 anaphylactic reactions (88.8% of which affected students) occurred in 607 U.S. schools responding out of 5683 schools surveyed. Eleven percent of schools reported at least one anaphylactic event, 2.1% reported two events, and 1.1% reported three or more events. Of the 851 events for which EAI use information was recorded, 75% were treated with EAI, 49% with stock EAI. Most of the anaphylactic events not treated with EAI were treated with antihistamines (White et al. 2015).

Data on stock epinephrine use are also available from the Chicago Public School District's Emergency Epinephrine Initiative to stock undesignated EAI in all 675 schools (DeSantiago-Cardenas et al. 2015). During the first year of the program (the 2012–2013 school year), 38 EAIs were administered, with most (92.1%) going to students. Over half (55%) of the usages were for food-induced reactions, with school nurses administering most (76.3%) of the EAIs. More recently, the state of Illinois released data from the 2014–2015 school year on the use of undesignated epinephrine at schools within the entire state (Illinois State Board of Education 2016). During that year, 59 schools reported a total of 65 administrations, of which 41 occurred within the Chicago Public School District.

Following the 2012 passage of legislation in Virginia to require local school boards to adopt policies to allow administration of epinephrine, the Virginia Department of Health and Virginia Department of Education conducted an epinephrine use survey of school divisions in the state (Virginia Department of Health 2014). Data were collected from 97% of school divisions, although it is unknown if all schools within divisions provided information. During the 2012–2013 school year, about 1.2 million students were enrolled in Virginia public schools. A total of 448 stock epinephrine doses were administered to students during that year, much more than the 166 doses of student-supplied epinephrine administered during that same time period. Most the stock epinephrine doses (336) were administered to students with no previously known allergies. An additional 28 doses of stock epinephrine were administered to staff or visitors during that school year.

Data collected from the use of stock EAI in schools have confirmed that students may not be aware that they have an allergy until they have an anaphylactic event at school. The EpiPen4Schools pilot study found 21.9% of individuals with an anaphylactic event had no known allergy. The Chicago study found 55% of EAI use was for first-time anaphylactic events (DeSantiago-Cardenas et al. 2015). The more recent 2014–2015 school year study in the state of Illinois found that of students (and staff) receiving 65 undesignated EAI administrations in schools, more than half (58.5%) had not previously been diagnosed with a severe allergy (Illinois State Board of Education 2016).

4.3.3   Stock Epinephrine in Schools Outside of the U.S.

Besides the U.S., other countries have completed or are beginning initiatives to stock undesignated epinephrine in schools:

- Many Australian territories require or recommend that schools stock EAIs (Vale et al. 2015).
- Europe has been slower to consider stock EAI in schools (Muraro et al. 2014), although, in the UK, the Anaphylaxis Campaign is working with other agencies to allow schools there to maintain stock EAIs (Anaphylaxis Campaign 2016).
- Following the school lunchtime death of an 11-year-old girl with a dairy allergy, Japan is also trying to develop policies such tragedies from occurring (The Japan Times 2013).
- Food Allergy Canada is also working to supply some schools with stock epinephrine autoinjectors (FAC 2015). The 2014 food allergy death of 14-year-old Canadian Caroline Lorette led to the formation of the Sweet Caroline Foundation, which advocates for stock epinephrine in schools.

4.3.4   Remaining Challenges for Epinephrine Availability in Schools

Although tremendous progress (and likely numerous lives) has been saved by current initiatives to stock epinephrine in schools, some challenges still exist. A general lack of awareness of food allergies and their treatment remains an issue at many schools. School staff may think calling 911 is sufficient without realizing how important for epinephrine administration to occur as fast as possible. As discussed later in Sect. 4.7 of this chapter, not all emergency rigs and personnel will carry or be able to administer epinephrine, further delaying life-saving treatment.

Work remains in bringing stock epinephrine into more schools. Several recent studies found that low socioeconomic status schools may not have stock EAI (Shah et al. 2014; Love et al. 2016). However, some very large school districts (Los Angeles, Chicago) are requiring stock EAIs at all schools. Cost is a big factor for many school districts and is becoming a bigger issue. As of August 2016, a two-pack of epinephrine autoinjectors costs $500 or more when previously they were only $100 (Popken 2016). The expiration date for epinephrine autoinjectors is typically 18 months (usually 14–16 months after a school receives it), so EAIs must be replaced almost every school year. Another way cost affects epinephrine availability is subtler: parents may assume schools have stock EAIs and therefore forego spending their own money on an EAI for their child.

4.4   Epinephrine Availability in Other U.S. Public Locations

Anaphylaxis can occur in other public locations besides schools. Making epinephrine available at public entities such as recreation camps, colleges, day care facilities, youth sports leagues, amusement parks, restaurants, places of employment, and sports arenas has lagged a bit behind school initiatives, partly because it has been difficult to get good statistics on anaphylaxis reactions that occur in some of these locations. Many states (30 as of March 2017) now have laws allowing public entities to obtain undesignated EAI (FARE 2015b). Some of these laws provide exemption from liability or provide training requirements for EAI use. FARE along with many other organizations assisted in the development of many of these laws by working with legislatures, supplying model language for bills, and providing data on epinephrine usage. Additionally, FARE identified local volunteers to be advocates to testify regarding the need for stock epinephrine and encouraged citizens to call their legislators to express their opinions.

4.5   Restaurant Preparedness

Several factors converge to make restaurants a place where food allergy reactions may occur. Food workers often lack the knowledge to prevent or treat anaphylaxis (Dupuis et al. 2016). A lack of appropriate training programs and incentives for restaurants to train staff coupled with high turnover rates for employees all hamper food allergy training at restaurants. Restaurants may not have space to properly segregate food allergens when preparing special meals for customers with allergies, while the need to prepare and serve many different meals quickly may cause mistakes to be made. Importantly, food allergic reactions may occur within minutes to hours after ingesting an allergen (Saleh-Langenberg et al. 2016; Sampson et al. 1992), meaning anaphylaxis can occur while someone is still at a restaurant. Chapters 9–11 of this book discuss in more detail these and other food allergy challenges that restaurants face.

FARE has been involved with several initiatives to improve restaurant management of food allergens. As described in Chap. 9 of this book, FARE collaborated with the National Restaurant Association to create the ServSafe® food allergy training program (which was modeled on existing foodborne pathogen training) for restaurant workers. FARE has also advocated for required food allergy training to improve food allergy awareness for restaurant workers in various states.

Apart from state initiatives (discussed in Chap. 3), several cities such as New York City and St. Paul, Minnesota also have requirements for food allergy information posters to be posted in food service establishments. New York City's required food allergy sign, which is available in eight different languages, lists the eight most common food allergens, describes how a server should react to a customer with a food allergy, and lists steps the kitchen staff and servers can take to prevent cross-contact (NYC Health 2016).

FARE has also created a food allergy awareness poster (available by free download) for restaurants to post in order to improve staff awareness of food allergies (Fig. 4.3).

As discussed in the other chapters discussing food allergen management control in restaurants (Chaps. 9–12), challenges still exist. Restaurant food allergy management laws differ greatly between states, making the development of general strategies for universal adoption difficult. "Food allergy friendly" remains hard to describe. As an example, the Food Allergy Awareness Act required the state of Massachusetts to work with the Massachusetts Restaurant Association (MRA) and FARE's predecessor, FAAN, to develop a program for restaurants to be designated as "Food Allergy Friendly" (FAF) and to maintain a listing of restaurants receiving that designation on the Department's website. This law was passed in 2009, but the listing has not yet been created, in part because of the difficulty in defining what "Food Allergy Friendly" means.

FARE has been involved in efforts to elevate food allergens as a food safety issue. As discussed in Chap. 3, the Food Code is an excellent example of the increased attention food allergens are receiving, with the 2013 Food Code providing guidance on minimizing cross-contact with allergens during food preparation for the first time (FDA 2016).

4.6   College and University Preparedness

A food allergic student who leaves home to attend college or university encounters new risks related to their food allergy. They may have roommates who are unfamiliar with food allergies who may not realize the dangers that even a single peanut can mean to certain individuals. The student may obtain a job or take a course that

Fig. 4.3 Food allergy awareness poster (provided by FARE)

involves working with laboratory animals whose diet may include dried milk powders or wheat products. Students will be on their own when it comes to making meal choices; their parents will not be there to guide their selections, and they will dine out in new places, whether in dining halls and cafeterias, restaurants, or other facilities.

Adolescents and young adults may also be more likely to engage in risky behaviors, and this behavior carries over into how they may manage their food allergies. One study (Sampson et al. 2006) found that a majority of food allergic individuals in this age group (13–21 years) has purposely ingested an unsafe food. In addition, a U.K. study found that food allergic individuals in this general age group (second and third decades of life) have the highest rates of hospital admissions and fatalities from food-induced anaphylaxis (Turner et al. 2015).

FARE has developed a College Food Allergy Program whose goals are to build allergen preparedness beyond dining halls and help colleges and universities "develop comprehensive, uniform food allergy management policies" (FARE 2016d).

In developing the program, FARE held two summits to meet with representatives from U.S. colleges and universities and other stakeholders to discuss food allergy policies that could be implemented by high-education institutions to improve food safety and quality of life for students with food allergies. These summits were attended by representatives from more than 50 colleges and others, who worked together with FARE to generate a document, "Pilot Guidelines for Managing Food Allergies in Higher Education" (FARE 2015a) as well as student and parent educational materials. These guidelines have been considered in development of food allergen management policies and programs for college and university dining services as described in Chap. 14 of this book.

Using knowledge gained from the summits, FARE piloted its College Food Allergy Program in 2014 with 12 colleges participating. The program has now expanded, with 15 additional colleges and universities added as of June 2016. Components of this program, which is free to participating institutions, are shown in Fig. 4.4.

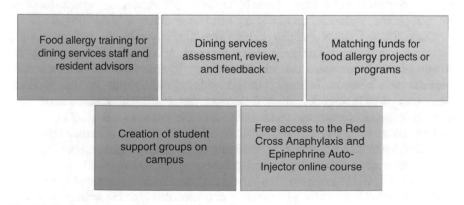

Fig. 4.4 Components of FARE's College Food Allergy Program

College and universities are public entities, so under the public entity laws discussed in Sect. 4.4 of this chapter, they may keep and administer stock epinephrine if their state permits it. Other state and university initiatives have worked together to make stock epinephrine available on college campuses in New Jersey (Princeton University), Indiana, and New Hampshire.

4.7 Emergency Treatment of Anaphylaxis

Anaphylaxis is not currently a rare event. In the U.S., about 30,000 emergency department visits and hospitalizations occur each year due to anaphylaxis (Ma et al. 2014). Anaphylaxis may occur in 1.6% of the population at some point during their lifetime (Simons et al. 2015). Recent studies have shown that both hospital visits and pediatric emergency room visits for anaphylaxis have been increasing (Dyer et al. 2015; Simons et al. 2015).

Quick treatment of anaphylaxis is important. The median time from food allergen exposure to death in anaphylaxis deaths is only 30 min (Pumphrey 2000). Late administration (>30 min) of epinephrine is associated with a poor outcome (McLean-Tooke et al. 2003). However, only one-third of adults who went to the emergency room for a food allergy reaction arrived within 1 h of exposure to the food allergen (Banerji et al. 2010). Improving the ability to treat anaphylaxis where it occurs (schools, restaurants, and other public locations, as discussed earlier in this chapter) should improve treatment outcomes and remains a priority. However, emergency responders will still have the first opportunity to treat a patient undergoing anaphylaxis in many cases.

Approximately 0.2–0.9% of emergency medical service transports are for allergic complaints (Manivannan et al. 2014). Surprisingly, however, a significant number of emergency rigs (ambulances) do not carry injectable epinephrine or may not have personnel trained in its use (FARE 2016b). In some cases, legal restrictions are in place that interfere with the ability of emergency rigs to carry or the ability of emergency personnel to deliver epinephrine. Only 34% of states have no restrictions on epinephrine availability on ambulances and other emergency rigs (Fig. 4.5).

Currently, ten states mandate epinephrine autoinjectors on emergency rigs, including Connecticut, Louisiana, Massachusetts, Maryland, Michigan, Oregon, Rhode Island, Tennessee, Utah, and Washington. Illinois previously required epinephrine autoinjectors, but now permits vials of epinephrine in lieu of EAIs (Amendment to Emergency Medical Services Act 2016). Other states leave EAI availability on emergency rigs under local control, have no official list of equipment and provisions for emergency rigs, or do not address epinephrine in their laws.

In addition to laws that restrict the availability of EAI on emergency rigs, only about 50% of states allow emergency medical technicians (EMTs) to use an EAI (Fig. 4.6). In some states, EMTs are allowed to use rig EAI or to assist a patient with their own EAI. Other states defer to local government or do not clearly define what is allowed.

Fig. 4.5 Restrictions on
epinephrine availability on
emergency rigs (adapted
from FARE)

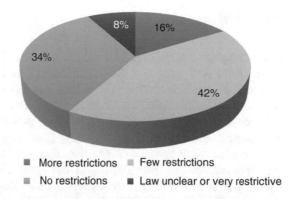

■ More restrictions   ■ Few restrictions
■ No restrictions   ■ Law unclear or very restrictive

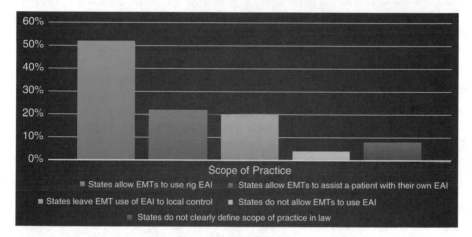

Fig. 4.6 State restrictions on EAI use by emergency medical technicians (adapted from FARE)

Even when present in emergency rigs, epinephrine may not be used. One study found that of 103 emergency transports of patients with allergic complaints, only 15 patients received epinephrine (Manivannan et al. 2014). Epinephrine was not used consistently across the study, and the authors concluded that standardized guidelines for anaphylaxis management were needed (Manivannan et al. 2014).

To address these known gaps in epinephrine availability and consistent utilization on emergency rigs, FARE is presenting information on these gaps at emergency medical conferences and other professional development opportunities to educate EMTs, emergency management services offices, and others about the proper diagnosis and treatment of anaphylaxis as well as the need for epinephrine on all emergency rigs. Long-term FARE goals include ensuring every emergency rig responding to a 911 call is stocked with epinephrine and have at least one crew member trained to administer it, and ensuring that every emergency stocks epinephrine and uses it as the first line treatment for anaphylaxis.

4.8   Airlines

An anaphylactic reaction is an emergency at any time, in any location. However, when such an event occurs during an airplane flight, the challenges in getting prompt medical treatment including epinephrine can have tragic consequences. Serious anaphylactic events during commercial flights are fortunately rare events, with only 265 occurrences (out of 11,920 medical emergencies) reported on 7.2 million global passenger flights between 2008 and 2010 (Peterson et al. 2013). Allergic reactions do, however, occur at a much higher rate during air travel; 9% of patients with peanut, tree nut, or seed allergies report having an allergic reaction while in flight (Comstock et al. 2008). One recent survey found that 33% of travelers who experienced an allergic reaction while flying reported symptoms consistent with anaphylaxis; however, only 10% of those with reactions received epinephrine (Greenhawt et al. 2009). These results were consistent with those of earlier reports suggesting a low rate of epinephrine use during anaphylactic events on airplanes (Sicherer et al. 1999; Comstock et al. 2008).

Although epinephrine is available within an onboard medical kit on most flights for use in other medical emergencies such as cardiovascular conditions, it is generally in the form of an ampule for intravenous administration and not explicitly labeled for use in anaphylaxis. Airlines allow travelers to carry self-injectable epinephrine, but as discussed earlier, sometimes an anaphylactic event is the first time an individual experiences an allergic reaction and would not therefore have an EAI with them.

Preventing reactions by eliminating common allergenic foods on airlines would seem a reasonable approach. Peanuts, for example, are both a common food allergen and a popular airline snack. However, rules banning certain foods would be very difficult to enforce; even if the airline didn't allow peanuts on flights, it would be exceedingly difficult to prevent passengers from bringing them aboard. In addition, the Federal Aviation Administration (FAA) will not allow a ban on foods in flight until a study is completed that demonstrates that such a ban would be beneficial. Since such a study would be virtually impossible to conduct.

A 2015 bill, Airline Access to Emergency Epinephrine Act S. 1972 (Airline Access to Emergency Epinephrine Act 2015), was introduced in the Senate and then referred to the Committee on Commerce, Science, and Transportation. Among other measures, the bill would have required airlines to carry epinephrine autoinjectors (EAIs), to replace them upon expiration or use, and to train crewmembers to recognize the symptoms of an acute allergic reaction and to administer EAI. The bill also instructed the FAA and individual airlines to clarify that the 1:1000 epinephrine ampules that are currently included in emergency medical kits may be used for the treatment of anaphylaxis. This bill, however, did not progress in the Senate (Airline Access to Emergency Epinephrine Act 2015).

A second bill, the FAA Reauthorization Act of 2016, was introduced by the Senate (Federal Aviation Administration Reauthorization Act of 2016 2016). This bill would have evaluated the medical equipment and supplies required on airline flights and ensure that required medical items meet the emergency medical needs of children, including consideration of an epinephrine autoinjector.

FARE worked closely to help craft language on both bills. Unfortunately, neither bill was successful as of the time of this writing.

4.9   Summary and Conclusions

Many challenges face those with food allergies and their families, including lack of public awareness, complicated and inconsistent labeling and policies, and social isolation and stress. An additional and critical problem these individuals face is the inability to receive prompt treatment with epinephrine in the case of an emergency in certain public locations and situations. Progress, catalyzed in some cases by tragedy, has been made in addressing these issues in recent years, largely through the passage of legislation that improves food allergy training and awareness and improves availability to emergency epinephrine. Advocacy organizations have worked with the government to improve epinephrine availability in public schools and other public locations. In addition, the efforts of groups like FARE have facilitated the passage of legislation that requiring food allergy training and awareness tools for restaurant personnel. Challenges still exist and more work is underway. Food allergy awareness programs at colleges and universities are being implemented, and efforts to ensure all emergency rigs and airlines carry epinephrine (and are trained and able to use it) are initiatives that are currently being addressed by FARE among other organizations. The use of appropriate legislation to drive these initiatives continues to be a powerful way to enact change within states and throughout the nation.

Acknowledgements   The author would like to thank several colleagues within and outside of FARE who offered their expertise in compiling this information at the 2015 conference at which it was presented: Nancy Gregory, Anne Thompson, Steve Gendel, Steve Taylor, Linda Temple, Jenny Kleiman Dowd, Michael Pistiner, as well as Wendy Bedale who elegantly organized the presentation into this chapter.

References

Agarwal, N.S., and J.E. Yu. 2015. Assessment of food allergy knowledge in NYC elementary school teachers. *Journal of Allergy and Clinical Immunology.* 135: AB138–AB138.
Airline Access to Emergency Epinephrine Act. 2015. Act # S. 1972, 114th Congress. Available online at: https://www.congress.gov/bill/114th-congress/senate-bill/1972. Accessed 1 June 2017.
Akeson, N., A. Worth, and A. Sheikh. 2007. The psychosocial impact of anaphylaxis on young people and their parents. *Clinical and Experimental Allergy.* 37: 1213–1220.
Amendment to Emergency Medical Services Act. 2016. 99th Illinois General Assembly.
Anaphylaxis Campaign. 2016. Adrenaline auto-injectors to be held in all schools - Campaign. Available at: http://www.anaphylaxis.org.uk/campaigning/generic-pen-campaign/. Accessed 1 September 2016.

Banerji, A., S.A. Rudders, B. Corel, A.M. Garth, S. Clark, and C.A. Camargo. 2010. Repeat epinephrine treatments for food-related allergic reactions that present to the emergency department. *Allergy and Asthma Proceedings*. 31: 308–316.

Branum, A.M., and S.L. Lukacs. 2008. Food allergy among US children: Trends in prevalence and hospitalizations. *NCHS Data Brief*(10): 1–8.

———. 2009. Food allergy among children in the United States. *Pediatrics*. 124: 1549–1555.

CDC (Centers for Disease Control and Prevention). 2013. *Voluntary guidelines for managing food allergies in schools and early care and education programs*. Washington, DC: US Department of Health and Human Services.

Chafen, J.J.S., S.J. Newberry, M.A. Riedl, D.M. Bravata, M. Maglione, M.J. Suttorp, V. Sundaram, N.M. Paige, A. Towfigh, B.J. Hulley, and P.G. Shekelle. 2010. Diagnosing and managing common food allergies: A systematic review. *JAMA - Journal of the American Medical Association*. 303: 1848–1856.

Chokshi, N.Y., D. Patel, and C.M. Davis. 2015. Long-term increase in epinephrine availability associated with school nurse training in food allergy. *Journal of Allergy and Clinical Immunology: In Practice*. 3: 128–130.

Comstock, S.S., R. DeMera, L.C. Vega, E.J. Boren, S. Deane, L.A.D. Haapanen, and S.S. Teuber. 2008. Allergic reactions to peanuts, tree nuts, and seeds aboard commercial airliners. *Annals of Allergy Asthma & Immunology*. 101: 51–56.

Dahlman, G 2013. The promise of the School Access to Emergency Epinephrine Act. FARE Blog. Available at: https://blog.foodallergy.org/2013/12/11/the-promise-of-the-school-access-to-emergency-epinephrine-act. Accessed 1 June 2017.

DeMuth, K.A., and A.M. Fitzpatrick. 2011. Epinephrine autoinjector availability among children with food allergy. *Allergy and Asthma Proceedings*. 32: 295–300.

DeSantiago-Cardenas, L., V. Rivkina, S.A. Whyte, B.C. Harvey-Gintoft, B.J. Bunning, and R.S. Gupta. 2015. Emergency epinephrine use for food allergy reactions in Chicago Public Schools. *American Journal of Preventive Medicine*. 48: 170–173.

Dudley, L.S., M.I. Mansour, and M.A. Merlin. 2015. Epinephrine for anaphylaxis: Underutilized and unavailable. *Western Journal of Emergency Medicine*. 16: 385–387.

Dupuis, R., Z. Meisel, D. Grande, E. Strupp, S. Kounaves, A. Graves, R. Frasso, and C.C. Cannuscio. 2016. Food allergy management among restaurant workers in a large US city. *Food Control*. 63: 147–157.

Dyer, A.A., C.H. Lau, T.L. Smith, B.M. Smith, and R.S. Gupta. 2015. Pediatric emergency department visits and hospitalizations due to food-induced anaphylaxis in Illinois. *Annals of Allergy Asthma and Immunology*. 115: 56–62.

Farbman, K.S., and K.A. Michelson. 2016. Anaphylaxis in children. *Current Opinion in Pediatrics*. 28: 294–297.

Federal Aviation Administration Reauthorization Act of 2016. 2016. Act # S.2658, 114th Congress. Available at: https://www.congress.gov/bill/114th-congress/senate-bill/2658. Accessed 1 June 2017.

Food Allergy Canada. 2015. Sussex, New Brunswick unveils new stock epinephrine program. Available at: http://foodallergycanada.ca/2015/09/sussex-new-brunswick-unveils-new-stock-epinephrine-program/. Accessed 5 December 2016.

FARE (Food Allergy Research & Education). 2015a. Pilot guidelines for managing food allergies in higher education. Avaiable at: http://www.foodallergy.org/file/college-pilot-guidelines.pdf. Accessed 9 September 2016.

———. 2015b. Public access to epinephrine. Avaiable at: http://www.foodallergy.org/advocacy/advocacy-priorities/access-to-epinephrine/public-access-to-epinephrine. Accessed 22 March 2016.

———. 2016a. About FARE. Avaiable at: http://www.foodallergy.org/about. Accessed 8 September 2016.

———. 2016b. EMTs and epinephrine. Avaiable at: http://www.foodallergy.org/advocacy/ems. Accessed 20 January 2017.

———. 2016c. Epinephrine at school. Avaiable at: http://www.foodallergy.org/advocacy/epinephrine-at-school. Accessed 5 May 2016.

———. 2016d. Resources for colleges and universities. Avaiable at: http://www.foodallergy.org/resources/colleges-universities. Accessed 9 September 2016.

———. 2016e. School guidelines. Avaiable at: http://www.foodallergy.org/laws-and-regulations/guidelines-for-schools. Accessed 5 December 2016.

FDA (Food and Drug Administration). 2016. FDA Food Code. Available at: http://www.fda.gov/Food/GuidanceRegulation/RetailFoodProtection/FoodCode/default.htm. Accessed 20 January 2017.

Greenhawt, M.J., M.S. McMorris, and T.J. Furlong. 2009. Self-reported allergic reactions to peanut and tree nuts occurring on commercial airlines. *Journal of Allergy and Clinical Immunology.* 124: 598–599.

Gupta, R.S., J.S. Kim, J.A. Barnathan, L.B. Amsden, L.S. Tummala, and J.L. Holl. 2008. Food allergy knowledge, attitudes and beliefs: Focus groups of parents, physicians and the general public. *BMC Pediatrics.* 8: 10.

Gupta, R.S., E.E. Springston, M.R. Warrier, B. Smith, R. Kumar, J. Pongracic, and J.L. Holl. 2011. The prevalence, severity, and distribution of childhood food allergy in the United States. *Pediatrics.* 128: E9–E17.

Gupta, R., D. Holdford, L. Bilaver, A. Dyer, J.L. Holl, and D. Meltzer. 2013. The economic impact of childhood food allergy in the United States. *JAMA Pediatrics.* 167: 1026–1031.

Hefle, S.L., T.J. Furlong, L. Niemann, H. Lemon-Mule, S. Sicherer, and S.L. Taylor. 2007. Consumer attitudes and risks associated with packaged foods having advisory labeling regarding the presence of peanuts. *Journal of Allergy and Clinical Immunology.* 120: 171–176.

Illinois State Board of Education. 2016. Report of use of undesignated epinephrine school year 2014-15. Avaiable at: http://www.isbe.net/pdf/school_health/epinephrine-use-report14-15.pdf. Accessed 5 December 2016.

Kemp, S.F., R.F. Lockey, F.E.R. Simons, and World Allergy Organization ad hoc Committee on Epinephrine in Anaphylaxis. 2008. Epinephrine: The drug of choice for anaphylaxis. *A statement of the World Allergy Organization. Allergy.* 63: 1061–1070.

Love, M.A., M. Breeden, K. Dack, A. Milner, A.C. Rorie, and S.A. Gierer. 2016. A law is not enough: Geographical disparities in stock epinephrine access in Kansas. *Journal of Allergy and Clinical Immunology.* 137: AB56–AB56.

Ma, L., T.M. Danoff, and L. Borish. 2014. Case fatality and population mortality associated with anaphylaxis in the United States. *Journal of Allergy and Clinical Immunology.* 133: 1075–1083.

Manivannan, V., R.J. Hyde, D.G. Hankins, M.F. Bellolio, M.G. Fedko, W.W. Decker, and R.L. Campbell. 2014. Epinephrine use and outcomes in anaphylaxis patients transported by emergency medical services. *American Journal of Emergency Medicine.* 32: 1097–1102.

McIntyre, C.L., A.H. Sheetz, C.R. Carroll, and M.C. Young. 2005. Administration of epinephrine for life-threatening allergic reactions in school settings. *Pediatrics.* 116: 1134–1140.

McLean-Tooke, A.P.C., C.A. Bethune, A.C. Fay, and G.P. Spickett. 2003. Adrenaline in the treatment of anaphylaxis: What is the evidence? *British Medical Journal.* 327 (7427): 1332–1335.

Muraro, A., I. Agache, A. Clark, A. Sheikh, G. Roberts, C.A. Akdis, L.M. Borrego, J. Higgs, J.O. Hourihane, P. Jorgensen, A. Mazon, D. Parmigiani, M. Said, S. Schnadt, H. van Os-Medendorp, B.J. Vlieg-Boerstra, and M. Wickman. 2014. EAACI food allergy and anaphylaxis guidelines: Managing patients with food allergy in the community. *Allergy.* 69: 1046–1057.

Mylan Specialty L.P 2016. Epinephrine in schools. Available at: https://www.epipen.com/en/hcp/for-health-care-partners/for-school-nurses. Accessed 3 April 2017.

Nowak-Wegrzyn, A., M.K. Conover-Walker, and R.A. Wood. 2001. Food-allergic reactions in schools and preschools. *Archives of Pediatrics and Adolescent Medicine.* 155: 790–795.

NYC Health. 2016. Food allergy poster samples. Available at: https://www1.nyc.gov/assets/doh/downloads/pdf/rii/allergy-poster-samples.pdf. Accessed 6 December 2016.

Odhav, A., C.E. Ciaccio, M. Serota, and P.J. Dowling. 2015. Barriers to treatment with epinephrine for anaphylaxis by school nurses. *Journal of Allergy and Clinical Immunology.* 135: AB211–AB211.

Patel, D.A., D.A. Holdford, E. Edwards, and N.V. Carroll. 2011. Estimating the economic burden of food-induced allergic reactions and anaphylaxis in the United States. *Journal of Allergy and Clinical Immunology*. 128: 110–U187.

Peterson, D.C., C. Martin-Gill, F.X. Guyette, A.Z. Tobias, C.E. McCarthy, S.T. Harrington, T.R. Delbridge, and D.M. Yealy. 2013. Outcomes of medical emergencies on commercial airline flights. *New England Journal of Medicine*. 368: 2075–2083.

Pieretti, M.M., D. Chung, R. Pacenza, T. Slotkin, and S.H. Sicherer. 2009. Audit of manufactured products: Use of allergen advisory labels and identification of labeling ambiguities. *Journal of Allergy and Clinical Immunology*. 124: 337–341.

Popken, B. 2016. EpiPen price hike has parents of kids with allergies scrambling ahead of school year. NBC News. Available at: http://www.nbcnews.com/business/economy/epipen-price-hike-has-parents-kids-allergies-scrambling-ahead-school-n633071. Accessed 5 December 2016.

Portnoy, J.M., and J. Shroba. 2014. Managing food allergies in schools. *Current Allergy and Asthma Reports*. 14: 7.

Pumphrey, R.S.H. 2000. Lessons for management of anaphylaxis from a study of fatal reactions. *Clinical and Experimental Allergy*. 30: 1144–1150.

Saleh-Langenberg, J., B.M.J. Flokstra-de Blok, N. AlAgla, B.J. Kollen, and A.E.J. Dubois. 2016. Late reactions in food-allergic children and adolescents after double-blind, placebo-controlled food challenges. *Allergy*. 71: 1069–1073.

Sampson, H.A., L. Mendelson, and J.P. Rosen. 1992. Fatal and near-fatal anaphylactic reactions to food in children and adolescents. *New England Journal of Medicine*. 327: 380–384.

Sampson, M.A., A. Munoz-Furlong, and S.H. Sicherer. 2006. Risk-taking and coping strategies of adolescents and young adults with food allergy. *Journal of Allergy and Clinical Immunology*. 117: 1440–1445.

Shah, S.S., C.L. Parker, E.O. Smith, and C.M. Davis. 2014. Disparity in the availability of injectable epinephrine in a large, diverse US school district. *Journal of Allergy and Clinical Immunology-In Practice* 2: 288–293.e1.

Sheikh, A., Y.A. Shehata, S.G.A. Brown, and F.E.R. Simons. 2009. Adrenaline for the treatment of anaphylaxis: Cochrane systematic review. *Allergy*. 64: 204–212.

Sicherer, S.H., T.J. Furlong, J. DeSimone, and H.A. Sampson. 1999. Self-reported allergic reactions to peanut on commercial airliners. *Journal of Allergy and Clinical Immunology*. 104: 186–189.

———. 2001. The US peanut and tree nut allergy registry: Characteristics of reactions in schools and day care. *Journal of Pediatrics*. 138: 560–565.

Simons, F.E.R., M. Ebisawa, M. Sanchez-Borges, B.Y. Thong, M. Worm, L.K. Tanno, R.F. Lockey, Y.M. El-Gamal, S.G.A. Brown, H.S. Park, and A. Sheikh. 2015. 2015 update of the evidence base: World Allergy Organization anaphylaxis guidelines. *World Allergy Organization Journal*. 8: 16.

Song, T.T., and P. Lieberman. 2015. Epinephrine in anaphylaxis: Doubt no more. *Current Opinion in Allergy and Clinical Immunology*. 15: 323–328.

Spina, J.L., C.L. McIntyre, and J.A. Pulcini. 2012. An intervention to increase high school students' compliance with carrying auto-injectable epinephrine: A MASNRN study. *Journal of School Nursing*. 28: 230–237.

The Japan Times. 2013. Avoiding food allergy tragedies. *The Japan Times*. Available at: http://www.japantimes.co.jp/opinion/2013/05/20/editorials/avoiding-food-allergy-tragedies/. Accessed 23 June 2017.

Turner, P.J., M.H. Gowland, V. Sharma, D. Ierodiakonou, N. Harper, T. Garcez, R. Pumphrey, and R.J. Boyle. 2015. Increase in anaphylaxis-related hospitalizations but no increase in fatalities: An analysis of United Kingdom national anaphylaxis data, 1992-2012. *Journal of Allergy and Clinical Immunology*. 135: 956–963.

Vale, S., J. Smith, M. Said, R.J. Mullins, and R. Loh. 2015. ASCIA guidelines for prevention of anaphylaxis in schools, pre-schools and childcare: 2015 update. *Journal of Paediatrics and Child Health*. 51: 949–954.

Virginia Department of Health. 2014. Epinephrine report.

Walkner, M., C. Warren, and R.S. Gupta. 2015. Quality of life in food allergy patients and their families. *Pediatric Clinics of North America.* 62: 1453–1461.

White, M.V., D. Goss, K. Hollis, K. Millar, S. Silvia, P.H. Siegel, M.E. Bennett, M.J. Wooddell, and S.L. Hogue. 2016. Anaphylaxis triggers and treatments by grade level and staff training: Findings from the EPIPEN4SCHOOLS pilot survey. *Pediatric Allergy Immunology and Pulmonology.* 29: 80–85.

White, M.V., S.L. Hogue, M.E. Bennett, D. Goss, K. Millar, K. Hollis, P.H. Siegel, R.A. Wolf, M.J. Wooddell, and S. Silvia. 2015. EpiPen4Schools pilot survey: Occurrence of anaphylaxis, triggers, and epinephrine administration in a US school setting. *Allergy and Asthma Proceedings.* 36: 306–312.

Zurzolo, G.A., M. de Courten, J. Koplin, M.L. Mathai, and K.J. Allen. 2016. Is advising food allergic patients to avoid food with precautionary allergen labelling out of date? *Current Opinion in Allergy and Clinical Immunology.* 16: 272–277.

Chapter 5
Food Allergen Recalls: The Past as Prologue

Steven M. Gendel

5.1 Introduction

The wide-spread application of Hazard Analysis and Critical Control Points (HACCP) principles to food safety systems is intended to protect public health by preventing potential problems before they happen rather than responding to problems after they occur. This approach is considered particularly efficacious for hazards that occur irregularly or nonuniformly in foods. For example, facilities can reduce risk of microbial hazards in food more effectively by applying a validated microbial kill step during production than by final product testing.

This preventive approach is embedded in food safety regulations in the U.S. such as the FDA juice (FDA 2003) and seafood (FDA 1999) HACCP regulations, the USDA meat HACCP regulation (FSIS 2015b), and the FDA preventive controls regulations (for both human and animal foods) (FDA 2016). HACCP has also been adopted by international bodies such as Codex Alimentarius Commission (2009) and is the basis for private audit standards such as GFSI (2016) and ISO (2005).

Despite the apparently forward-looking approach implicit in HACCP and preventive controls, all of these schemes rely on a description of the past to determine which hazards need to be controlled in the future. This is particularly evident in the hazard analysis. The first step in a hazard analysis for a particular food and facility is to identify those hazards that are, for example, "known or reasonably foreseeable" (FDA 2016) or "reasonably likely" (FSIS 2015b). To be "known" means that a hazard has been associated with the particular food or facility in the past. To be "reasonably likely" or "reasonably foreseeable" suggests that this association was not a one-time

S.M. Gendel (✉)
Division of Food Allergens, IEH Laboratories and Consulting Group,
Lake Forest Park, WA, USA
e-mail: steven.gendel@iehinc.com

© Springer International Publishing AG 2018
T.-J. Fu et al. (eds.), *Food Allergens*, Food Microbiology and Food Safety,
DOI 10.1007/978-3-319-66586-3_5

event. There is also an implicit assumption that the future will be similar to the past, and that controlling those hazards that occurred in the past will result in safe food in the future. Therefore, in order to conduct a complete hazard analysis, information on the association between foods and hazards over time must be available.

Information on the association between foods and pathogens is available from multiple sources. For example, information is made available by CDC on food-borne outbreaks both through an online database (CDC 2016) and scientific publications (CDC 2013). There are also many papers in the technical literature describing microbiological surveys of foods obtained in different regions of the world. The CDC makes some information available about illnesses associated with some chemical hazards, while FDA tracks a defined set of chemical hazards in the Total Diet Study (Egan et al. 2002). However, there are no comparable sources of information on food allergens. In addition, there are only a few published studies on the presence of undeclared allergens in foods, and these generally focus on the use of allergen advisory statements (Zurzolo et al. 2013; Crotty and Taylor 2010).

Food allergens are a unique public health issue in that the primary risk mitigation tool available to food manufacturers is the provision of allergen information on food labels. Food allergic consumers rely on labels to be clear, complete, and accurate. This means that all allergens that are intended to be present should be declared and that no undeclared allergens should be present. Allergen information may be inaccurate or incomplete because a label fails to declare an allergen that is present as an ingredient, an unintended allergen is present through cross-contact or mislabeling, or an allergen is declared that is not present (Gendel and Zhu 2013). Both of the first two situations can create significant public health risks for food allergic consumers who need to practice avoidance to prevent potentially life-threatening reactions. The declaration of an allergen that is not present in a product is not in itself a threat to public health, but usually occurs in conjunction with labeling or packing errors that result in the presence of an undeclared allergen in another product. Any of these situations can lead to a food recall in the U.S.

The only available information that can be used to support prevention-based food allergen control is from allergen-related recalls. Many food safety regulatory agencies provide some recall information as a service to consumers, and this information can also be used to support an allergen control hazard analysis. To demonstrate how this can be done, data on allergen-related food recalls in the first half of calendar year 2015 were obtained from three food regulatory agencies in North American (FDA, FSIS, and CFIA) and from the Rapid Alert System for Food and Feed (RASFF) in Europe. Significant gaps in the publically available information were also identified.

5.2 Data Sources

Information on allergen-related recalls was obtained online from each of the three North American food regulatory agencies: FDA, USDA/FSIS, and CFIA. Data on recalls during the period from January 1 to June 30, 2015 for FDA regulated products were obtained from the agency "Recalls and Safety Alert Archives" web page at http://www.fda.gov/Safety/Recalls/ArchiveRecalls/default.htm. Data on recalls during this time period for USDA/FSIS regulated products was obtained from the "Recall Case Archives" at http://www.fsis.usda.gov/wps/portal/fsis/topics/recalls-and-public-health-alerts/recall-case-archive and the Current Recalls and Alerts page at http://www.fsis.usda.gov/wps/portal/fsis/topics/recalls-and-public-health-alerts/current-recalls-and-alerts. Data on recalls for Canadian foods were obtained from the CFIA web site at http://www.inspection.gc.ca/about-the-cfia/newsroom/food-recall-warnings/complete-listing/eng/1351519587174/.

Data from each of these sources were used to identify the food and allergen(s) involved in each recall and the recall classification. Although each of these agencies operates under different statutory authorities and has responsibility for different segments of the food supply, they all use a three-class system to categorize recalls. Class I recalls are situations where the public health risk is high; Class II recalls are situations where the public health risk is lower; and Class III recalls are situations where direct harm is unlikely or where there is a technical violation (Table 5.1).

Data from the RASFF database were obtained using the online portal at http://ec.europa.eu/food/safety/rasff/index_en.htm by searching for "allergens" as the hazard and using the date range January 1–June 30, 2015. Information in the RASFF database does not correspond to that provided by the three North American agencies, but consists of alert notifications, information notifications, and border rejection notifications from government agencies in the participating member countries. These alerts may or may not be the result of, or triggers for, recalls or market withdrawals.

Table 5.1 Standards for classification of food recalls in North America

Agency	Class I	Class II	Class III
FDA[a]	"… reasonable probability that … a violative product will cause serious adverse health consequences or death"	"… a violative product may cause temporary or medically reversible adverse health consequences or where the probability of serious adverse health consequences is remote"	"… a violative product is not likely to cause adverse health consequences"
FSIS[b]	"… a *reasonable* probability that eating the food will cause health problems or death"	"… a *remote* probability of adverse health consequences from eating the food"	"… eating the food will not cause adverse health consequences"
CFIA[c]	"… a violative product will cause serious adverse health consequences or death"	"… a violative product may cause temporary adverse health consequences or where the probability of serious adverse health consequences is remote"	"… a violative product is not likely to cause any adverse health consequences"

[a]FDA (2011)
[b]FSIS (2015a)
[c]CFIA (2014)

5.3   Data Analysis

Information was obtained for 530 recalls from the three North American regulatory agencies for the period from January 1 to June 30, 2015. Table 5.2 shows the allergen-related recalls for each agency as well as the number of food allergen recalls in each recall class. On average, 51% of all North American food recalls were related to undeclared allergens and 52% of these were Class I recalls. The CFIA had a larger proportion of Class III recalls (24%) than either of the U.S. agencies (0.6% for FDA and 0% for FSIS).

Table 5.3 shows the food allergens most commonly involved in FDA recalls during this period. There were a number of recalls linked to the detection of peanut proteins in imported cumin during this period, and these recalls were tabulated separately (FDA 2015). Among the FDA recalls, 36 were linked to finding peanut protein in imported cumin, leading to recalls for multiple products that contained this cumin. Of these, 18 were Class I, 17 were Class II, and 1 was Class III.

Table 5.4 shows the most common allergens involved in FSIS recalls during this period. In this case, all of the 15 peanut recalls were linked to the cumin situation. Of these, 14 were Class I and 1 was Class II.

Table 5.5 shows the most common allergens for the CFIA recalls. In Canada during this time, five of the six peanut recalls were related to cumin. Of these, four were Class I and one was Class III.

During the same period there were 1514 notices added to the RASFF database, of which 58 (4%) were allergen-related. Table 5.6 shows the allergens involved in RASFF notices during this time period. The information available in the public portal did not make it possible to determine whether any of these allergen notices were related to cumin.

Table 5.2 North American food allergen recalls during the first half of 2015

Agency	Total recalls	Allergen recalls[a]	Class I[b]	Class II[b]	Class III[b]
FDA	258	178 (69%)	115 (65%)	62 (35%)	1 (0.6%)
FSIS	95	43 (45%)	26 (60%)	17 (40%)	0
CFIA	177	71 (40%)	23 (32%)	31 (44%)	17 (24%)

[a]Number and % of total recalls
[b]Number and % of allergen recalls for that agency

Table 5.3 The most common food allergens in FDA recalls

Allergen	Recalls
Peanut/Cumin	36
Peanut/Other	28
Milk	31
Soy	19
Tree Nuts	18
Egg	15

Table 5.4 The most common food allergens in FSIS recalls

Allergens	Recalls
Peanut/Cumin	15
Peanut/Other	0
Soy	14
Wheat	9
Egg	6
Milk	5

Table 5.5 The most common food allergens in CFIA recalls

Allergens	Recalls
Milk	22
Mustard	19
Wheat	11
Sesame	7
Peanut/Cumin	5
Peanut/Other	1

Table 5.6 Food allergens in RASFF (EU) notifications

Allergens	Notifications
Milk	13
Wheat/Gluten	12
Tree Nuts	12
Soy	9
Egg	7
Peanut	5

5.4 Discussion

Although these data present a limited snap-shot of food allergen recalls, there are several interesting insights that can be derived from this information. First, it is clear that undeclared food allergens continue to be a significant public health and industry problem, at least in North America. As has been seen in other studies, about half of all food recalls are caused by undeclared food allergens (Gendel and Zhu 2013; Gendel et al. 2014). There is no sign that this high rate of allergen-related recalls is decreasing, despite the improved awareness for allergen control by the food industry. Given the high cost to the company (or companies) involved in a food recall, the cost to consumers who have reactions to undeclared allergens, and the loss of public confidence in the ability of companies to control allergens, it can be seen that these recalls are a huge economic and physiological burden to society.

Second, the classifications of the cluster of cumin-related recalls that affected products regulated by all three North American agencies show an apparent difference in the way that these agencies interpret the metrics used to determine recall classification (Table 5.1). Both the FSIS and CFIA classified almost all of these recalls as Class I (93% and 80%, respectively) while only 50% of the FDA recalls were Class I. Although none of these agencies has publically described how these classifications were determined, it appears that the FDA may have considered consumer exposure to peanut protein in final food products to a greater extent than did the other agencies. Understanding the factors considered by FDA, and their relative weights, would be helpful to manufacturers who need to determine when allergens are "hazards that require a preventive control" under the Preventive Controls for Human Foods rule that was published in September of 2015.

Third, it is clear that, unlike pathogens, Class II recalls are a significant source of risk for allergic consumers. On average, 40% of the North American allergen-related recalls were Class II (see Table 5.1 and Gendel and Zhu (2013) for information on allergen recall classification). This is particularly important because the standard for reporting problems to the FDA Reportable Food Registry (RFR) is the same as for a Class I recall (Gendel et al. 2014). This means that information on food allergen risks is incomplete in the RFR database because not all recalled foods are included in the RFR despite a public health risk.

When the peanut in cumin recalls are disregarded, undeclared milk is the most frequent cause of allergen-related recalls. About 20% of all the non-cumin-related recalls on North America were linked to undeclared milk, and this was the only non-peanut food allergen to cause recalls or notifications for all four agencies. This may reflect the widespread use of various milk derivatives in food formulations.

It is not possible to directly compare the number of notifications listed in the RASFF database to the recall information available from FDA, FSIA, and CFIA. The RASFF database includes data about a variety of problems, including border rejections, which are not recalls (RASFF 2015). Some notifications, for example the "information notifications," describe situations where "the risk is not considered serious or the product is not on the market at the time of notification." These situations would probably not lead to a recall in North America. In addition, this database is effectively a secondary resource because it only contains information submitted by the regulatory authorities in the participating countries. Nevertheless, it is worth noting that allergen-related problems are much less frequent in this database than they are in the recall databases from North America.

Integrating recall information across agencies or the whole food supply is difficult for several reasons. There are major differences between agencies in the content, format, and terminology in the publically available food recall data sets. The North American agencies are primarily concerned with providing short-term information to consumers as press releases. There is seldom any follow up after the initial press release (such as what corrective actions have been implemented) unless there is a need to expand a recall. Each agency makes recall information available through a different online "front end," and the structure and content of the recall notices can change from one recall to the next even in a single agency. The focus on consumer communication, and legal constraints, means that information on recalls of bulk products and products in business-to-business commerce (including products provided to food service operations) is not available.

Although these data provide valuable insights into concerns related to allergen control, what is arguably the most important information—the root cause of a recall—is not made public by any food safety agency. In a few cases, such as with peanut protein in cumin, a combination of the information in multiple press releases with other news sources provides some insight into the background of the recall. However, even in those few cases, this information does not identify the fundamental root cause. Information on root cause is becoming increasingly critical because it can be difficult to ensure that appropriate allergen controls are in place and are being properly monitored without understanding how allergen control systems fail.

5.5 Conclusion

The eventual success of regulatory preventive food safety systems will depend on a thorough understanding of the history of food-related public health problems. Although more information is currently available on the historical prevalence of microbial food safety, similar information is limited for food allergens. The use of data on allergen-related recalls could help to fill the data gap, but full realization of this potential will depend on the availability of more complete and well-structured data from food safety regulatory agencies.

There is also a clear need for a standardized approach to reporting recall data. Standardizing the specific information contained in each recall notification, the terminology used among the agencies, and providing access to these data in a format that supports automated analysis would be important steps in allowing the use of advanced data analysis to identify the causes of allergen-related recalls and protect the health of food allergic consumers.

Understanding the past is critical for protecting foods and consumers in the future, and this understanding is particularly difficult to obtain for food allergens. Hopefully, the widespread adoption of prevention- and risk-based food safety regulations will catalyze a change leading to increased data accessibility.

References

CDC (Centers for Disease Control and Prevention). 2013. Surveillance for foodborne disease outbreaks - United States, 2009–2010. *Morbidity and Mortality Weekly Report.* 62: 41–47.
———. 2016. Foodborne outbreak online database (FOOD Tool). Available at: http://www.cdc.gov/foodsafety/fdoss/data/food.html. Accessed 26 July 2016.
CFIA (Canadian Food Inspection Agency). 2014. Recall plans – Distributor's guide. Available at: http://www.inspection.gc.ca/food/safe-food-production-systems/food-recall-and-emergency-response/distributors-guide/eng/1376400892829/1376401519986#a4.2. Accessed 26 July 2016.
Codex Alimentarius Commission. 2009. Food hygiene basic texts. World Health Organization, Food and Agricultural Organization of the United Nations, Rome. Available at: http://www.fao.org/docrep/012/a1552e/a1552e00.htm. Accessed 23 June 2017.
Crotty, M., and S. Taylor. 2010. Letter to the editor: Risks associated with foods having advisory milk labeling. *Journal of Allergy and Clinical Immunology.* 125: 935–937.

Egan, S.K., S.S. Tao, J.A. Pennington, and P.M. Bolger. 2002. US Food and Drug Administration's total diet study: Intake of nutritional and toxic elements, 1991-96. *Food Additives and Contaminants.* 19: 103–125.

FDA (Food and Drug Administration). 1999. Guidance for industry: HACCP regulation for fish and fishery products; Questions and answers for guidance to facilitate the implementation of a HACCP system in seafood processing. Available at: http://www.fda.gov/Food/GuidanceRegulation/GuidanceDocumentsRegulatoryInformation/Seafood/ucm176892.htm. Accessed 26 July 2016.

———. 2003. Guidance for industry: The juice HACCP regulation - Questions and answers. Available at: http://www.fda.gov/food/guidanceregulation/guidancedocumentsregulatoryinformation/ucm072602.htm. Accessed 26 July 2016.

———. 2011. Recalls background and definitions. Available at: http://www.fda.gov/Safety/Recalls/IndustryGuidance/ucm129337.htm. Accessed July 26, 2106.

———. 2015. FDA consumer advice on products containing ground cumin with undeclared peanuts. Available at: http://www.fda.gov/food/recallsoutbreaksemergencies/safetyalertsadvisories/ucm434274.htm. Accessed 26 July 2016.

———. 2016. FDA Food Safety Modernization Act. Available at: http://www.fda.gov/Food/GuidanceRegulation/FSMA/default.htm. Accessed 26 July 2016.

USDA-FSIS (United States Department of Agriculture – Food Safety and Inspection Services). 2015a. FSIS food recalls. Available at: http://www.fsis.usda.gov/wps/portal/fsis/topics/food-safety-education/get-answers/food-safety-fact-sheets/production-and-inspection/fsis-food-recalls/fsis-food-recalls. Accessed 26 July 2016.

———. 2015b. HACCP. Available at: http://www.fsis.usda.gov/wps/portal/fsis/topics/regulatory-compliance/haccp. Accessed 26 July 2016.

Gendel, S.M., and J. Zhu. 2013. Analysis of U.S. Food and Drug Administration food allergen recalls after implementation of the Food Allergen Labeling and Consumer Protection Act. *Journal of Food Protection.* 76: 1933–1938.

Gendel, S.M., J. Zhu, N. Nolan, and K. Gombas. 2014. Learning from FDA food allergen recalls and reportable foods. *Food Safety Magazine.* 20: 46–52.

GFSI (Global Food Safety Initiative). 2016. GFSI benchmarking requirements: Guidance document. Available at: http://www.mygfsi.com/schemes-certification/benchmarking/gfsi-guidance-document.html. Accessed 26 July 2015.

ISO (International Organization for Standardization). 2005. ISO22000:2005. Available at: http://www.iso.org/iso/home/store/catalogue_tc/catalogue_detail.htm?csnumber=35466. Accessed 26 July 2016.

RASFF (Rapid Alert System for Food and Feed). 2015. RASFF preliminary annual report 2015. Available at: http://ec.europa.eu/food/safety/rasff/index_en.htm. Accessed 26 July 2016.

Zurzolo, G.A., J.J. Koplin, M.L. Mathai, S.L. Taylor, D. Tey, and K.J. Allen. 2013. Foods with precautionary allergen labeling in Australia rarely contain detectable allergen. *Journal of Allergy and Clinical Immunology: In Practice.* 4: 401–403.

Chapter 6
Allergen Management: Ensuring What Is in the Product Is on the Package

Susan Estes

6.1 Introduction

Food allergen management has become a very important food safety program over the past 20 years. Prior to the early 1990s, little was known about the allergenic properties of food, and there was little done to control allergens by the food industry. From that time, awareness in all sectors of the industry has grown, including the manufacturing industry as well as regulatory agencies. As this awareness has grown, more mistakes are being identified. Some may think that the food industry is doing a poorer job at managing these risks. But a better explanation may be that mistakes that are being made were never identified before allergen controls were put in place and actually happened much more frequently. Allergen management-related recalls have increased by 41% from 2007 to 2012 (Gendel and Zhu 2013). According to the authors of that publication, 67% of recalls with a known root cause were the result of failures in label control. These failures include wrong product, wrong package (such as old packaging), and an incorrect formula version for the product in the package. These findings highlight the need for effective label management programs to control this significant food safety hazard.

There are many components of a food allergen management program. Some of these are:

1. Product development strategies
2. Label/Packaging controls
3. Allergen cross-contact controls
4. Verification programs
5. Training

S. Estes (✉)
Global Food Safety, PepsiCo, Inc. (Retired), Barrington, IL, USA
e-mail: Estessa44@gmail.com

© Springer International Publishing AG 2018
T.-J. Fu et al. (eds.), *Food Allergens*, Food Microbiology and Food Safety,
DOI 10.1007/978-3-319-66586-3_6

These program components can be simply segregated into two objectives: (1) ensuring accurate labeling of allergens in ingredients used in the formula and (2) keeping allergens not in the formula out of the product. While these objectives seem easy to achieve, they can be very challenging. There are many control points. All are critical in an allergen control program. A failure at one point may cause the whole system to fail. This chapter focuses on the second objective, but it is important to look holistically at how both objectives are related, and the fact that all need to be effective to protect the food allergic individual.

An effective allergen label management program requires putting in place controls at every stage of the packaging supply chain, from package development to application/filling of the final package. This supply chain includes the development and design phase, when ingredient evaluation, formulation, supplier ingredient and packaging controls must be considered. In production, label reviews must be conducted at receipt, during storage and at point of use. Training is critical, along with record keeping of packaging use, and corrective actions implemented, if a failure occurs. This chapter covers risk assessment and controls for both the design phase and the operations phase.

6.2 Design Phase

6.2.1 Formulation

The first step to correctly label a product is to understand the allergen content of the ingredients that are going to be used in the formula. While this might seem simple, some important questions and discussions may be necessary. Many ingredients have multiple components and the composition of some, such as flavoring systems, may not be obvious. Allergens should only be added when necessary for the functionality or characterization of the product. There are situations when using an ingredient is required, such as peanuts in peanut butter. In other instances, an alternative, such as a corn starch carrier for a flavor to replace wheat starch, is a preferred option as the wheat allergen is now removed.

6.2.1.1 Package Design Team

The coordinated effort of many departments is required to develop a comprehensive and legal package label. The design team will be a cross-functional team, including:

– Product, Process and Package Development (providing information on the product and process)
– Regulatory Affairs (to ensure legality and compliance)
– Quality, Food Safety and Supplier Quality (to provide the risk assessments detailed below).

All must work as a team with ingredient suppliers to identify allergen containing ingredients and manage the risk to the business in order to prevent a costly recall.

6.2.1.2 Understanding the Ingredients in Your Product

To have an accurate understanding of the ingredients in your product, the first step is to provide a questionnaire to the ingredient supplier. This questionnaire should include a number of inquiries about the allergen content of the proposed material. Some of these questions may include:

- Is an allergen present?
- Is the allergen intentionally formulated or present due to cross-contact?
- Name of the allergen component in the ingredient.
- Form of the allergen component in the ingredient.
- Concentration of allergen present in the ingredient.
- Allergen concentration determined by what method? (analytical or calculated)
- This question will help in the risk assessment (analytical data is always preferred)
- Are other allergens present in the supplier's facility?

While this is not an all-inclusive list of questions, it should give a good start and can be revised as more experience is gained. If the capability exists, an electronic survey form may be the most convenient way to gather this information. Formulation software solutions do exist to facilitate this approach. If this solution also connects with a formulation tool, the generation of a label can also be an output. But if budget is a concern, an email (or paper) gathering of the information would work just as well. If a form is used, it could look something like the example in Fig. 6.1.

It is important to remember that a form must be requested for each ingredient, each supplier, and all suppliers' manufacturing locations. In some situations, there could be some differences in minor ingredients from different suppliers. Additionally, even within a single supplier, the allergen profile of ingredients produced at various manufacturing locations and lines may vary. It is therefore necessary to have a completed form from each supplier's manufacturing location. A completed example is included in Fig. 6.1.

6.2.1.3 Risk Assessment of Supplier/Manufacturing Site Information

Once the information has been received from the supplier, an experienced employee should review the information. This will usually be done by the Food Safety or Regulatory Affairs member of the Design Team. The reviewer should understand ingredients, allergen content implications, and regulatory considerations. Once the review has been completed, some changes may be necessary by the product developer and/or supplier. Some items that the design team may want to consider:

Food Ingredients, Inc.

Ingredient Name:___Peanut Butter Chip #12345_ Supplier:_Chips, Inc.
Supplier Location: ____Chicago, IL_____

Allergen	Contains	Formulated	Cross Contact	Name of Component	Form of Component	Concentration	Method	In Facility
Milk								
Egg								
Soybean								
Wheat								
Tree Nut								
Peanut	X	Yes	No	Peanut Butter	Particulate	10 ppm	Neogen Veratox	No
Fish								
Shellfish								

Fig. 6.1 An example of a supplier ingredient survey

- Working with ingredient suppliers to remove unnecessary allergens
- Avoiding the use of allergenic minor ingredients that do not add to the product's profile or function

Another factor that must be considered is the current line configuration at the site that will make the new product. The team must understand the current existing allergen profile of the line or facility that is going to make the product to determine if the introduction of an allergen into the process has implications for food safety, production practices, or label design for existing products. They must also have an understanding of current products made on the line to be used and any indications of cross-contact that would impact the label of the new product.

All documentation around labeling decisions and assessments should be maintained for future reference. This type of historical information can be helpful in future decisions or if challenged.

6.2.1.4   Risk Assessment of Ingredient Suppliers

It will be necessary to understand the allergen controls in place at the ingredient manufacturer's location. This review will help you to understand the level of risk to the ingredient for cross-contact and also the quality of the information provided, and is usually done by the Supply Quality member of the team. Unfortunately, the level of understanding around allergen management controls is not the same across the industry. There may be an opportunity to educate and partner with the supplier to increase their understanding of allergen risks and improve their programs. It is also necessary to confirm that the supplier cannot change ingredient composition or the processing environment without first notifying the company. It should be made clear that the expectation is that if a change is made that impacts the information previously shared, the information should be revised and resubmitted.

6.2.2 Developing the Label

6.2.2.1 The Ingredient Statement

Once these considerations have been addressed, the next step is to determine how the ingredients will be listed in the ingredient statement of the product. Many countries have regulations that specify which allergens must be declared on the package, and in what format. In the U.S., for example, the Food Allergen Labeling and Consumer Protection Act (FALCPA) (FDA 2004) requires the top eight allergens be labeled in plain English on the product package. Two format options are allowed: the allergen is listed in the body of the ingredient statement in common language, i.e., whey (milk). Another option is to use a statement at the end of the ingredient line, such as: Contains milk. Both of these options can be used, but the two cannot be in conflict with one another. A limitation of current labeling regulations is that the presence of any amount of allergen requires its presence on the label. All food products containing allergens, even with small amounts, are required to declare their presence. While there is much debate around "how much is too much" in the allergy community, there are data to show that some very low levels of allergens (ppm or mg amounts), may not provoke an allergic response in sensitive individuals (Taylor et al. 2010). The establishment of allergen thresholds may eliminate the labeling of ingredients with very small amounts of allergens, thus expanding food choices for these patients who already have a limited diet.

Once the appropriate text for the ingredient information has been determined, it will be necessary to make sure that this information makes it to the label. An internal approval process for graphics prototypes should be in place to ensure that this happens. There should be an accountable party to review and sign off on each component of the label to make sure that the ingredient information that was developed is correctly captured on the graphic. It is also important to remember to review multipack and/or external carton labels to ensure consistency with the individual items within the packs or cartons, as the external package labels are being created.

A point to remember in the development cycle: Products made specifically for R&D purposes that will be consumed by anyone, even employees, should have proper allergen labeling indicating any allergens present in the product.

6.2.2.2 Precautionary Statements

FALCPA does not address allergen cross-contact, or the inadvertent incorporation of an allergen into a food. The food industry uses advisory or precautionary labeling when there is a risk of cross-contact during food manufacture. This would apply to the inadvertent presence of allergens in incoming ingredients as well as the actual impact of the line used in the manufacturing of the product. As precautionary labeling is voluntary, the food industry is left to determine, by company, when and how it is used. The approach that companies are taking varies widely. Some will label a product with a precautionary statement if an allergen is present in a facility, even if there is no risk of that allergen getting into the product. This approach unnecessarily limits the choices of food allergic consumers. More importantly, foods that really do carry a risk, and are labeled to warn of this risk, will be ignored by the food allergic consumer.

A best practice approach includes the premise that precautionary labeling should be used with great caution, and only as an exception rather than a default approach. It should only be used on rare occasions, and never as a substitute for good cleaning practices. Precautionary labeling should be used in situations when the presence of the allergen is documented, sporadic, uncontrollable, and potentially hazardous. Some examples of evidence would be finished product testing which indicates the significant presence of an unformulated allergen after a complete and thorough cleaning has been conducted. This is an indication that the equipment can't be adequately cleaned, and there is a risk to the consumer. Precautionary labeling should only be used when there is a true risk to someone with the allergy in question, and not just a mechanism to potentially protect the company from liability. If all companies would adhere to this approach, the labeling to warn of inadvertent allergen content would become meaningful and not be ignored by allergic consumers. Establishment of allergen thresholds could potentially standardize use of advisory labeling so that if levels above those considered a public health concern are found in a product, the product would be labeled appropriately.

6.2.2.3 Products Produced for Export

U.S. manufactured products intended for an international market will need to be properly labeled according to requirements of the country where the product will be sold. Regulatory requirements vary, so it is necessary to ensure that specific requirements for the country of sale are met. Some companies may proactively label allergens to accommodate frequently exported products (such as the labeling of sesame for Canada on U.S. products), for ease of label and line management.

6.2.3 Package Design: Appearance

6.2.3.1 Design Considerations to Minimize Operational Failures

As the marketing team is developing the design/layout of the package, there are some considerations that can be helpful in differentiating packaging between products. These differences can be helpful in making sure the right product gets into the right package. Differentiation can also help the consumer find the product they would like to purchase—a dual benefit.

One way to accomplish this is to have different color flags for different flavors. In the examples below (Figs. 6.2 and 6.3), different colors are used to designate different flavors of the same product.

While this approach is helpful, a better example is the one shown in Fig. 6.4. Different colors of packaging make the distinction between flavors much more apparent. It is helpful in the manufacturing plant to distinguish between flavors, but also helps the consumer find the flavor they want to buy.

6.2.3.2 Design Considerations for Enhanced Consumer Awareness

Another consideration is the physical design of the packaging when the allergen profile is changed. This is important when an allergen is being added to a product that did not previously contain the allergen. As we know, allergic patients and their caregivers are voracious label readers. But sometimes these individuals become comfortable with a product and may not check the package.

Fig. 6.2 Example of a differentiation of product flavors by flavor name color differences (Courtesy: PepsiCo, Inc.)

Fig. 6.3 Differentiation of title color for product flavor type (Courtesy: PepsiCo, Inc.)

Fig. 6.4 An example of color differentiation of flavors by using different bag colors for different flavors (Courtesy: PepsiCo, Inc.)

Fig. 6.5 Potential flags to allergic consumers that the product has changed and the ingredient listing should be checked (Courtesy: PepsiCo, Inc.)

To help these individuals be aware of a change to the product, a flag on the front panel stating "New Formula" or "New and Improved" can act as an indicator to encourage them to read the ingredients to see if a change has been made that may impact their ability to eat a product. Highlighting the allergens, in bold letters or some other fashion, may also be helpful (see Fig. 6.5).

6.3   Operations Phase

6.3.1   Label/Packaging Supplier Controls

6.3.1.1   Package Engraver

Ultimately, manufacturers are responsible for putting the correct product into the right packaging with a product label that matches the contents of the package. But ensuring the label information is translated to the package begins at the engraver of the die plates. The graphic layout is transferred from the design team to the engraver, who will manufacture the plates for printing. Controls must be in place here to make sure that the information is transferred correctly, that obsolete die plates are destroyed, and inventory tracking is implemented.

6.3.1.2 Packaging Manufacturer

A control point at the packaging supplier is to verify that the correct die plate is used to print the packaging. The product name and formulation version must be carefully managed and tracked. Additionally, there should be a process in place to make sure that labels/packages are not mixed with labels of other products or versions. Finally, there should be controls set in place to destroy old plates, old packaging, and any other outdated materials at the packaging manufacturer.

While these controls are the responsibility of the package supplier to execute, a process to audit their programs and procedures will give the food manufacturer assurance that these controls are effective.

6.3.2 Controls at Food Manufacturing Locations

6.3.2.1 Label/Packaging Review Upon Receipt

The next stage in the supply chain where label controls should be in place is at the product manufacturing location. As new labels/packaging arrive at the location, they should be reviewed for accuracy, to make sure the statements reflect the formulation that is going to be used to make the product. This can be done by a visual review or by an automated scan.

If labels are printed in-house, the manufacturing location should have a program in place to review each batch printed for accuracy. This may be done by manually checking the label for accuracy.

6.3.2.2 Label/Packaging Control During Storage/Staging for Production

The materials must then be inventoried. An inventory control system must be in place to proper storage and to account for destruction of packaging. The warehouse staff must be able to easily locate the appropriate packaging for the production run. This can be accomplished by using a warehouse management system that has a unique identifier for each material. This identifier can be scanned and documented prior to removal from the warehouse. The line attendant can also scan the identifier when the materials are placed on the line. When it is time to use the new materials, the old versions must be removed, defaced or preferably, destroyed. The old packaging should be sent to an approved landfill for destruction, with an affidavit of destruction sent after completion. This will eliminate the potential for the old material to be moved and used on the line.

Additional controls include storage of labels away from potential contact with ingredients or spills to prevent allergen exposure. Designated storage locations for packaging away from ingredients would be appropriate. It is also useful to establish a tracking system for removal of packaging from storage, placing packaging on the

Fig. 6.6 An example of how a color difference or text difference can be used for detection by a vision system (Courtesy: PepsiCo, Inc.)

line, and returning any partial batches to inventory. This system should include a record of the responsible party for each step for verification purposes. This can easily be accomplished using an online tracking system or simply a log sheet with initials of the designated responsible party.

6.3.2.3   Label Verification at the Point of Use

Controls at the point of line manufacturing are critical to correct labeling. Most mistakes are arguably caused by human error. Therefore, the best means for mini-mizing labeling errors is to make the manufacturing process as automated and mistake proof as possible.

Control solutions for label/package use can range in cost and complexity. Purchased equipment such as vision systems/UPC scanners can be costly but very effective. Vision systems can be calibrated to look for specific markers, unique to a specific package. Color differences, text differences, and other markers can be used to differentiate between packages. The machine must be properly calibrated to dif-ferentiate between packages. Some examples are demonstrated in Fig. 6.6.

The vision system can also be used to check the UPC code. This is the best way to make sure the right packaging for being run on the line. The UPC is unique, and will be a very specific control. It may also be helpful to track the formula version on the package as well. This also ensures that the version of the product package matches the version of the formula that is currently being manufactured (Fig. 6.7).

There are some inexpensive controls other than automated vision systems that can also be used. A simple visual comparator can be created. The tool can be as simple as a verified correct label with the allergens highlighted. This label can be

Fig. 6.7 An example of a designation of formula and version on a package to be used for tracking purposes (Courtesy: PepsiCo, Inc.)

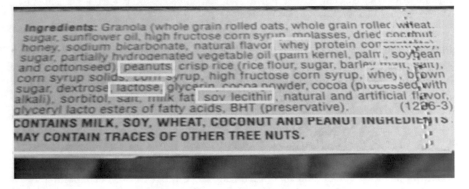

Fig. 6.8 By simply highlighting allergen containing ingredient, other labels can be compared to this example to make sure these ingredients have been correctly labeled

physically compared to any new labels to make sure that the highlighted ingredients have been captured. The example can focus primarily on allergens, allowing the line operator to quickly check that the allergens in the product are correctly reflected on the label when the packaging is brought to the line (Fig. 6.8). This is an easily executed quick check that may prevent a costly recall.

6.3.3   Staff Training

As with all components of an allergen management program, training is critical to correctly execute the packaging control portion of the program. General training should include allergen awareness training for all new employees and annual refresher training for all employees. Most importantly, specific documented training per job responsibilities is critical. For package controls, warehouse personnel should be trained on how to verify package accuracy upon receipt, the correct procedure for pulling of film/package material from stock, and stock rotation and the documentation required for both procedures. The line operator should be trained to verify the formulation and packaging match—the product and version being made are the same on the packaging. This check should be documented on the line paperwork.

6.4 Summary and Conclusions

As stated earlier, allergen related food safety incidents are the leading cause of recalls in the U.S. Therefore, allergen controls must be a part of the manufacturing site's food safety program. Regulations will dictate how controls are implemented and will be based on the standard level of care expected in the industry. All of the appropriate documentation and training must be in place to enable the food safety plan to be executed effectively.

There are controls that can be implemented throughout the supply chain to minimize the potential for allergen related packaging errors. These controls begin in the product development cycle. Controls in the package design and development process can also be implemented. Allergens in every supplier material used in production must be identified to ensure proper labeling. This step is critical for conveying what allergens are formulated into the product. Cross-contact allergens or unintentional allergens that pose a risk to the consumer must also be considered.

Once the allergen content of the product has been determined, controls are necessary to make sure this information is correctly included on the label via the ingredient statement. The review of package content prior to, during and after the actual production of the package is critical. There are tools available, both automated and manual, that can help with this process. These controls must be included in a food safety program and monitored effectively. Training is critical, both for awareness and specific job responsibilities.

These components are critical for a successful Allergen Management Plan to ensure that the manufacturer will succeed in the following key objective: Label what they add, and not add what they do not label!

References

FDA (U.S. Food and Drug Administration). 2004. Food Allergen Labeling and Consumer Protection Act of 2004 (Public Law 108-282, Title II). Available at: http://www.fda.gov/Food/GuidanceRegulation/GuidanceDocumentsRegulatoryInformation/Allergens/ucm106187.htm. Accessed 28 September 2016.

Gendel, S.M., and J. Zhu. 2013. Analysis of U.S. Food and Drug Administration food allergen recalls after implementation of the food allergen labeling and consumer protection act. *Journal of Food Protection*. 76: 1933–1938.

Taylor, S.L., D.A. Moneret-Vautrin, R.W.R. Crevel, D. Sheffield, M. Morisset, P. Dumont, B.C. Remington, and J.L. Baumert. 2010. Threshold dose for peanut: Risk characterization based upon diagnostic oral challenge of a series of 286 peanut-allergic individuals. *Food and Chemical Toxicology*. 48: 814–819.

Chapter 7
Allergen Management in Food Processing Operations: Keeping What Is Not on the Package Out of the Product

Timothy Adams

7.1 Introduction

Food allergens have become a tremendous concern for the packaged food industry. The increasing urgency that food manufacturers feel in controlling allergens is partly driven both by increasing numbers of people affected or aware of food allergies. A variety of data suggest an increasing prevalence of food allergies, especially in children (Sicherer 2011; Branum and Lukacs 2008). In addition, public awareness of food allergies, while still less than ideal, has also increased (Taylor and Hefle 2005).

Few companies in the food industry had food allergy management policies in place before 1990 (Taylor and Hefle 2005). Initial efforts to develop allergen management programs began in response to consumer demand, with the FDA becoming involved soon thereafter (Taylor and Hefle 2005). The FDA's Food Allergen Labeling and Consumer Protection Act of 2004 (FALCPA) was a pivotal law that was intended to make it easier for those with food allergies and their families to identify food products that contained allergens (FDA 2004). The law required food labels to highlight the presence of certain allergens within food products. For example, if lecithin was used in manufacturing a product, the label would have to make it clear that this product contained soy, either through a parenthetical statement after the ingredient ("lecithin (soy)") or by a separate "contains soy" statement.

FALCPA provided a way for consumers to know if **intentional** allergens were present in manufactured foods. The law, however, did not address the possibility of accidental allergen cross-contact, which is defined as when an allergen is accidentally transferred from a food containing an allergen to one that does not contain the allergen (FARE 2017). Precautionary advisory labeling (PAL) for

T. Adams (✉)
Kellogg Company, Battle Creek, MI 49017, USA
e-mail: tim.adams@kellogg.com

© Springer International Publishing AG 2018
T.-J. Fu et al. (eds.), *Food Allergens*, Food Microbiology and Food Safety,
DOI 10.1007/978-3-319-66586-3_7

allergens arose in the food industry to manage and communicate potential risk to consumers resulting from potential allergen carryover from other products manufactured on the same equipment or in the same facility (DunnGalvin et al. 2015). FALCPA did not require nor did it provide guidance regarding the use of precautionary advisory labeling (such as "may contain nuts") beyond stating that PAL should not replace good manufacturing practices and must be truthful and not misleading (FDA 2006).

FDA's landmark Food Safety Modernization Act (FSMA), which was signed into law in 2011, contains provisions to prevent accidental food allergen cross-contact in manufactured foods (FSMA 2011). FSMA has made food allergen control an explicit part of its preventive controls for human foods (PCHF) requirements, elevating food allergen management to a level comparable to that previously reserved for food pathogen management (FSPCA 2016). The preventive controls for human foods regulations specify that food allergens be considered when conducting the required hazard analysis for a manufactured product. Furthermore, the regulations require that controls be in place to prevent cross-contact and to ensure that labeling is performed correctly (FDA 2015).

Besides being logical in terms of public health, preventive controls for food allergens also make sense from a business perspective. Allergens represent a major source of food recalls both in the U.S. and internationally (Gendel and Zhu 2013; Bucchini et al. 2016). Recalls can be enormously expensive (Pozo and Schroeder 2016), especially class I food safety recalls (the category under which most allergen-related recalls fall). Recalls also generate bad publicity, increase insurance premiums, and in some cases may result in costly litigation for a company (Marler 2015).

To comply with the preventive controls requirements, to protect public health, and to protect the company's reputation and bottom line, an allergen management program is essential for food manufacturers. An allergen management program is the set of controls an organization uses to prevent unintended allergens from being present in products. The controls used may include activities related to virtually all parts of the manufacturing process, including ingredients and suppliers, labeling and packaging, product design, receiving and storage, engineering and system design, operational controls and scheduling, sanitation and changeover practices, equipment and facility maintenance, employee training, and consumer contact.

Each company's (and even each facility's) allergen management program will be unique and should receive input from representatives from manufacturing, quality, and regulatory, but also engineering, sanitation, customer service and other functional areas as appropriate. The allergen management program will be a living plan, to be reviewed and revised as needed when changes such as new ingredients or processes are introduced.

Two key objectives exist in every allergen management program:

- Correctly communicate the allergen content of the product to the consumer via labeling (discussed in Chap. 6)
- Keep allergens which are not on the label out of the product

This chapter focuses on the second of these objectives: keeping allergens that are not supposed to be in a product out of a product. The main components of an allergen management program are highlighted, with discussions on validation for allergen sanitation (Sect. 7.3.7) and allergen management program review (Sect. 7.3.8).

7.2   Sources of Allergen Cross-Contact

In order to conduct an allergen risk assessment, it is critical to think about and understand potential sources of allergen cross-contact. In the complicated and fast-paced environment of a food manufacturing facility, there are many opportunities for allergen cross-contact, some of which are shown in Fig. 7.1.

Some sources of allergen cross-contact in a manufacturing environment are easy to imagine. For example, if the same manufacturing equipment is used for both a cheese cracker and a plain cracker, inadequate sanitation procedures between runs could lead to milk allergen being present in the plain crackers. Similarly, manufacturing peanut candy on a line next to a line that makes a chocolate candy could result in a peanut fragment accidentally being introduced into the chocolate candy. Other sources of allergen cross-contact may be less obvious. If an employee eats a peanut butter sandwich at lunch, crumbs from that sandwich could fall from their beard while they are weighing ingredients for a manufacturing run. An ingredient supplier could make a small change that can have a big impact; for example, in one real-life example, chocolate was contaminated with peanut allergen when the cocoa bean supplier reused jute bags that had previously been used to ship peanuts (Taylor and Hefle 2005). Finally, human error can also result in allergen cross-contact: mistakes can happen during labeling, formulation, or in the manufacturing process that may result in the unexpected introduction of allergens.

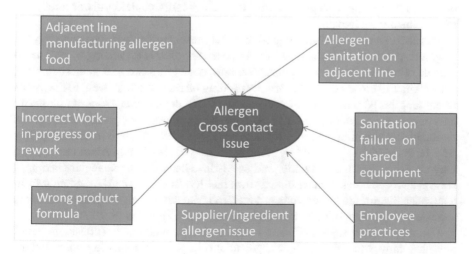

Fig. 7.1   Key sources of allergen cross-contact during food manufacturing activities

It is important to remain vigilant and to carefully consider the possible impact of each step (and every change) to a manufacturing process to ensure the potential for allergen cross-contact is considered and addressed within the allergen control program. Even companies that do not typically handle allergens may want to consider a program in case they inadvertently handle allergens because of mislabeling or contamination of ingredients used in their products.

7.3 Elements of Effective Allergen Control Programs

Chapter 6 of this book discusses two important controls that a manufacturing facility should use in their allergen control program: ingredients/supplier controls, and labeling/packaging controls. The subsections below in this chapter cover other key elements of allergen control that a manufacturing facility should consider employing. Additional resources are available to help with the development of an allergen control plan for your company (Deibel et al. 1997; Taylor et al. 2006; Jackson et al. 2008; GMA 2009; Ward et al. 2010; Gendel et al. 2013; FoodDrinkEurope 2013; FARRP 2017).

7.3.1 Product Design and Hazard Assessment

Careful product design can reduce food allergen risks in manufactured foods. First, consider the ingredients you use within all products manufactured in your facility. Do you really need that particular ingredient in that product? For example, if you can eliminate the use of milk from a product, you not only make that particularly product safe for those with milk allergies, but you also reduce the chances that other products manufactured in your facility will accidentally contain milk as a result of manufacturing carryover.

It is also critical to consider ingredients that present allergen risk because of your supplier's practices. For example, if your pecan supplier is expanding operations to include other tree nuts and peanuts, you now need to ensure that your supplier is controlling the cross-contact hazard. Depending on risk, you may want to find a different supplier who only produces pecans to reduce the risk that traces of non-pecan nuts or peanuts might be present in the ingredient you purchase from them. Another option in this situation would be to use advisory labeling. A "zero tolerance" policy may be considered "prudent" by some facilities (Hewitt 2012). Alternatively, you may choose to audit that supplier and assess first-hand the steps they are taking to control allergen cross-contact in their own facility. Any allergen risks coming from a supplier become risks for your product, too, especially since levels of allergens capable of triggering allergic reactions are very low (Taylor et al. 2002).

Formulation and supplier changes also need to be made with a consideration of how they may affect the allergenic potential of a product, including both obvious risks and more subtle ones. Formulation changes for one product can also impact the allergenic potential of other products made within the same manufacturing facility.

Considering the increasingly complex nature of the supply chains, a manufacturer needs to ensure (via supplier agreements, for example) that no allergens are introduced when changes are made by a supplier, and/or that any change made to an ingredient is communicated in advance so its impact can be assessed.

Hazard assessment, to include food allergens, is now required under FSMA's preventive controls for human foods regulations (FDA 2015). The hazard analysis process identifies where a hazard exists in a process and helps determine where controls are required to mitigate allergen risk. A cross-functional team approach should be adopted for the hazard analysis exercise, since the perspective from different functional groups may identify different hazards. For example, production workers may be better at identifying potential human errors that might occur during manufacturing, while those involved in receiving may know that certain ingredients are more prone to leaking during shipment.

The use of a process flow diagram is critical when conducting a hazard analysis in order to identify the point in a process where allergen is added and also where it could be inadvertently introduced. The facility HACCP team should verify the flow diagram and identify allergen hazards on the production floor during all normal operating conditions. A facility map may also be useful to help visualize product flow and potential sources of cross-contact. During the hazard analysis, consider options that allow the use of an alternate ingredient that is not an allergen, changing the addition point of the allergen within the manufacturing process (later is better to reduce the amount of equipment that will be contaminated), or obtain your ingredient from another source if that reduces the possibility of allergen cross-contact.

The hazard analysis needs to be reassessed when formulation or equipment changes are made. A new piece of equipment may be more difficult to clean, allowing allergen residue to remain and potentially contaminate subsequent runs on the same equipment. A formulation alteration might introduce an unexpected allergen because of a contaminant in a new ingredient, or the formulation alteration may require a change in cleaning methods in order to ensure removal of food residue from equipment.

The hazard analysis needs to consider the entire facility and its full capabilities. The ideal preventive control is obviously to have separate, dedicated facilities for products with different allergens. However, such measures are not generally feasible. Even when separate facilities exist, the potential for allergens to be inadvertently introduced to a facility within an ingredient needs to be considered. The hazard analysis should identify all of these gaps and develop programs, procedures, and controls to close them.

7.3.2   Receiving and Storage Practices

As discussed in Chap. 6, you should have certain supplier controls in place prior to receiving raw ingredients. For example, your company should have some agreements with your suppliers to ensure that you will be notified if any changes in the ingredient are made, or you may request a letter of guarantee that an ingredient does not contain a specific allergen.

It is critical that all food allergens that enter your facility be identified at the time they are received. The labels of all incoming materials need to be reviewed to ensure that the correct material was shipped. Shipping controls are also important: consider what else was shipped with the product, and what was previously shipped in the truck or container. In one real-life example, a bakery product was contaminated with peanut allergen because the wheat used to make the flour was transported in a rail car that had previously transported peanuts (Taylor and Hefle 2005).

Receiving controls should include inspection for possible damage and/or leaks. The facility should have a detailed response plan for any allergen spill, and housekeeping and other personnel should be aware of and trained on what to do in such an event. Containers with allergenic ingredients should be appropriately and prominently tagged or labeled (color-coding is helpful), then segregated from non-allergenic ingredients.

Dedicated storage rooms (and/or pallets or containers, as appropriate) for allergenic ingredients should be used whenever possible, with careful consideration of the volumes that will need to be stored, the types of packed ingredients, and how many different separation scenarios your facility will need. When storing allergenic and non-allergenic ingredients in the same area, the facility must establish practices to prevent allergen cross-contact such as routine inspection, covering of materials, removal of outer coverings, and avoiding storage of allergenic ingredients above non-allergenic ones.

7.3.3 Engineering and System Design Requirements

The sanitary design of equipment and tools used in your facility should be reviewed to ensure that they are accessible and cleanable to your company's definition of "allergen clean" (which starts with being "visibly clean"). Niche areas or "dead spots" in equipment may require remediation. The potential need for disassembly of equipment for cleaning and subsequent inspection should be considered. All surfaces of equipment that might be exposed to allergen need to be made of a material that is compatible with the sanitation processes that will be used and the frequency with which such processes will be performed.

The design of a facility and its infrastructure also plays an important role in avoiding allergen cross-contact. Air handling systems should be designed if possible to minimize allergen movement through air movement within the facility. Appropriate placement of floor drains can facilitate allergen sanitation by allowing equipment and facilities to be cleaned thoroughly with complete removal of waste water. Ceilings and overhead equipment should be considered as potential reservoirs for airborne allergens. Walls, curtains, and other methods of separation can be used to prevent allergen movement within a facility. Particular attention should be paid to cross-over points where one allergen zone intersects a different one, as these are locations where allergen cross-contact may be most likely to occur.

7.3.4 Operational Controls and Scheduling

Sometimes dedication and separation during manufacturing are not possible. In such circumstances, production scheduling represents the best tool that can be used to minimize allergen cross-contact between runs. Some basic rules are usually followed in these cases. First, allergenic foods are usually run last to minimize the chance that they can contaminate non-allergenic foods. Production runs should be grouped (or longer runs utilized) when possible to minimize the amount of changeover work required between non-allergenic and allergenic products.

Sufficient time must be scheduled for allergen sanitation and any associated sanitation verification activities. In some cases it may be more economical to relocate production lines to save money on cleaning and sanitation. The HACCP team needs to ensure that individuals involved in scheduling manufacturing runs which contain different allergens have the appropriate knowledge to ensure that the potential for allergen cross-contact is minimized. Tools such as the changeover matrix shown in Fig. 7.2 may be useful.

In this matrix, the type of sanitation regimen and appropriate verification procedures necessary for any two types of consecutive product runs are specified in the grid by different colors. For example, if the first run contains a milk allergen as an ingredient (A), then if the second run contains milk and egg (D), then a GMP-level sanitation process (to a visually clean level, for example) is adequate between runs (yellows squares), with no allergen verification needed. However, if the second run contains an egg allergen instead of the milk allergen (E), the sanitation procedure must be appropriate to remove the milk allergen, and verification that the milk allergen is absent is needed before the second run containing the milk allergen can proceed.

Preoperational controls include various activities in addition to scheduling that should occur before a manufacturing run to ensure that the correct product is made and packaged without allergenic cross-contact. A line start-up SOP should be in place which ensures the manufacturing line is cleared of all packaging and ingredients from previous runs. Formula and ingredient checks should be implemented to ensure that the correct recipe and ingredients are always used. Packaging checks should also be included to make sure the correct packaging is used. Any allergen containment requirements identified through plant risk assessments (dust collection, curtains, etc.) need to be verified to be in place. The use of designated equipment or allergen-designated (or non-allergen-designated) tools and utensils should also be verified as appropriate.

Controls during operation can take many different forms. For example, employee personal protective equipment (which might be color coded) could be employed for the allergen area, with movement to other production lines controlled (and in some cases, prohibited). The traffic patterns and flow of employees, ingredients, and packaging materials may need to be prescribed and monitored. Controls that prevent equipment or machines from being used before cleaning and sanitation (analogous to lock-out and tag-out systems used in EHS programs) may be helpful to ensure manufacturing runs only occur when all controls are verified to be in place. Any material that is held for work in progress (WIP), rework, or that is outside of process needs to be prominently and consistently labeled. Specific products where rework can be used should be appropriately identified. Labeling is important, as is traceability of rework.

Additional GMPs may include hygiene rules for employees, contractors, maintenance workers, visitors, etc. These rules may include not bringing allergenic personal foods into the plant, handwashing, use of gloves and other protective wear, and other rules that will prevent people from accidentally transferring allergen anywhere within the plant.

7.3.5   Maintenance Practices

Maintenance activities occurring within a plant may be unscheduled (due to an equipment failure) or routine procedures to keep equipment and the facility running smoothly. Controls and procedures should be in place to prevent allergen cross-contact during maintenance activities, including both in-process "hot work orders" and offline maintenance work. These controls should ensure tools, carts, and work areas are maintained and procedures are followed to minimize inadvertent allergen contamination. Training for maintenance personnel is also needed to ensure they understand the importance of preventing allergen cross-contact and are cognizant of this potential when they perform their duties. Off-line sanitation work can include a sanitation and inspection step to ensure allergen cross-contact risk is mitigated or controlled.

7.3.6   Sanitation and Changeover Practices

Procedures for sanitation and product/equipment changeover need to be clear and easy to follow.

Wet cleaning is the best approach for allergen removal as most protein allergens are soluble in water (Taylor et al. 2002). Unfortunately, it is not always possible to introduce water into a particular manufacturing environment (Jackson et al. 2008) (see also Chap. 8). Dry cleaning needs to be performed in a way that minimizes the formation of aerosols or airborne dust. Push-through cleaning using an inert, non-allergenic ingredient is another option, but may require careful validation to ensure that allergen is completely removed (Taylor et al. 2002).

Allergen sanitation standard operating procedures (SSOPs) need to be in place which consider the type of allergen that is present (paste vs. particulate; water vs. oil soluble; matrix composition), the concentration of the food allergen, and the type of food contact surface that is being treated. A changeover matrix such as that shown in Fig. 7.2 can be used to help define allergen sanitation requirements under different conditions. Unless separate, labeled tools and utensils are used, defined procedures

Allergen Change Over Matrix		Product After Change Over					
		(milk) A	(peanut) B	(none) C	(milk, egg) D	(egg) E	(none) F
Product Prior To Change Over	(milk) A		Allergen milk	Allergen milk	GMP	Allergen milk	Allergen milk
	(peanut) B	Allergen peanut		Allergen peanut	Allergen peanut	Allergen peanut	Allergen peanut
	(none) C	GMP	Push Through		GMP	GMP	Push Through
	(milk, egg) D	Allergen egg	Allergen milk, egg	Allergen milk, egg		Allergen milk	Allergen milk, egg
	(egg) E	Allergen egg	Allergen egg	Allergen egg	GMP		Allergen egg
	(none) F	GMP	Push Through	Push Through	GMP	Push Through	

Fig. 7.2 Example of a product scheduling changeover matrix

for sanitation tool cleaning and management should be developed. Signage can help reinforce compliance with these procedures.

Defined sanitation performance standards should be in place. "Allergen clean" begins with a visual inspection, but in some cases requires additional evidence and testing to ensure that a facility or equipment is suitable for the manufacture of a non-allergenic food product. Verification and validation plans for allergen sanitation are covered in the following section of this chapter.

7.3.7 Verification and Validation for Allergen Sanitation

Verification is defined as the routine monitoring of the defined process to show adherence to a standard. In other words, "We did what we said we would do." In the context of allergen cleaning, verification provides assurances that the allergen SSOP was performed as designed with all steps completed.

Allergen SSOPs often include checklists which specify the required sanitation activities. Such checklists may be combined or included with various other preoperations checklists. A visual inspection of all equipment prior to assembly is usually included in the SSOPs. Although allergen sanitation starts with a visually clean standard, in certain cases additional allergen specific testing may be warranted. A facility must decide their testing program and incorporate various tools including ATP, protein, or allergen-specific testing. The facility needs to be aware that typical sanitation verification tools like ATP and total protein do not give allergen specific information and use of a combination of both types of tests provides a more comprehensive approach (Jackson et al. 2008).

Testing can be done on equipment or on finished products. Sampling plans are critical and will differ depending on whether the potential allergen is a powder or a particulate (for example, a peanut). The sampling plan and expectations should be carefully defined in the allergen SSOP.

The frequency of verification testing needs to be carefully considered, with mechanisms in place to increase or decrease testing depending on the trends seen over time. Corrective actions for possible results of the testing should be prespecified. Employees should be trained on these procedures to ensure compliance when events requiring a corrective action occur.

In contrast to verification, validation is defined as the process and necessary steps to ensure the effectiveness of a particular process. In other words, validation demonstrates "what we did was effective." For an allergen SSOP, validation means demonstrating that the SSOP is effective in reducing an allergen hazard to an acceptable level.

The validation plan should prespecify which analytical methods will be used. Methods should be appropriate for your particular product. The validation plan should consider what will be tested (finished product or something else) and define a sampling plan with expectations. The frequency of conducting validation studies (initial and ongoing) should be specified, as should any conditions that would trigger the need for revalidation or additional validation testing. Training of all employees involved in validation, just as for verification, should also be included within the validation plan.

For both verification and validation testing plans, appropriate test methods for the allergen of interest need to be determined and a lab to perform testing identified. Knowledge of the allergen you are targeting is needed to identify the best test. It may be helpful to work with your testing lab to understand potential detection issues.

If multiple allergens are being used in a product, consider targeting a specific allergen to reduce testing time and expense. The choice of allergen might be based on the relative amounts of the allergens being used, the timing of when allergens are added within the manufacturing process, the difficulty of cleaning each allergen, and the form of the allergen (particulate, paste, powder). Multiple validation studies covering each allergen may be required to support some programs. For example, if a facility adds peanut paste at the end of the line, but also has a milk difference throughout an entire process, the facility will probably need to target both allergens to verify allergen sanitation and controls are effective.

When designing an allergen sanitation validation protocol, it is important to verify the presence of the allergen in or on the equipment and in the previous food to be cleaned and removed as positive controls. It is important to ensure that your analytical method works in both your "before-food" and "after-food" matrices with which you are working. Once such positive controls are collected, the allergen SSOP should be followed to a visibly clean standard.

The sampling plan for the allergen sanitation validation protocol should specify what equipment needs to be swabbed, and where on the equipment the swabs should be taken, and how much finished product should be tested. It may be also useful to include sampling intermediate product or final rinse water from CIP systems in your validation protocol. The sampling plan should also consider the type of allergen that might be present (particulate, powder, etc.) when determining how the finished product should be sampled and tested.

Test results need to be interpreted, with the interpretation of various results prospectively discussed in the validation protocol. In the event of a positive result, the steps that should be taken next should be prespecified in the protocol. Situations where repeat testing may be warranted should be discussed, and how to proceed following such testing (especially when conflicting results are obtained) should be part of the validation protocol.

7.3.8 Program Auditing

The allergen management program should be periodically reviewed. An internal program audit can help ensure program effectiveness in controlling allergen cross-contact risk.

Audits should be performed at a predetermined frequency by a cross-functional team. The review should carefully assess any changes made to the current ingredients, the introduction of any new products in the facility, and any changes to manufacturing or sanitation processes.

Consumer contact information should also be reviewed to identify any trends in complaints and ensure resolution of any corrective actions. It is important to consider where allergens could be the cause of any of your consumer contacts, even if allergens were not something the consumer had considered.

The audit should include a review of in-process records and procedures for line start-up, changeover, and product holds for any circumstances that could indicate allergen risk. In addition, sanitation inspection and testing records should be examined to look for potential trends or problems.

7.3.9 Consumer Contact and CAPA

Consumer contact systems should be designed to identify allergen issues. It is important that your consumer contact system is able to obtain information from contacts that allows for identification of the product. Identification information should include product lot number, where it was purchased, the customer's allergy history, and whether there is product remaining.

Your HACCP team should have procedures to review, investigate, and follow up on allergen-related consumer contacts. Possible trends that could indicate a potential allergen related issue such as potential allergic reactions or mixed/wrong foods should be monitored carefully. If a problem seems possible, investigate promptly and take steps (press releases, recalls, etc.) as necessary to prevent other customers from becoming ill.

A formalized approach including root cause analysis and corrective and preventative action (CAPA) should be in place to address any problems that are uncovered in these investigations. All investigations, analyses, and CAPA activities should be carefully documented and reviewed to verify that they were effective in addressing problems.

7.3.10 Allergen Training

Allergen training should be instituted for most if not all employees working at a food manufacturing facility. The training should include information on general allergen awareness as well as job-specific allergen information.

Job-specific training should be tailored to the responsibilities of a particular employee. For example, it is important for everyone involved in manufacturing to understand that the amount of allergen that can cause an allergic reaction can be exceedingly low and that the consequences of accidental exposure enormously great. This knowledge will help them comply with and potentially add improvements to the allergen management plan. Customer service representatives need to know the symptoms related to food allergy reactions in order to quickly spot trends that could indicate an allergen cross-contact problem in a product. Even visitors to the production floor may need some very brief allergen training to ensure they do not accidentally move into an allergen-free area after walking through a production area where an allergenic ingredient is being used.

Allergen training, as required for each employee, should be documented. Refresher training should be scheduled periodically. A matrix chart may be useful to track training requirements and designed frequency of training for each functional area or job title.

Finally, in addition to formal training, job aids such as signs can be useful within the manufacturing facility to remind employees of specific things they need to do (or not do) to prevent allergen cross-contact while they are working. This type of signage not only reinforces what has been learned in training, but also helps demonstrate to the employee the importance the company places on preventing allergen cross-contact. Training resources are available, include the USDA's Countertop Food Safety Training Program (USDA 2017).

7.4 Summary and Conclusions

A comprehensive allergen management program for a food manufacturing facility should use a hazard/risk assessment process to help identify possible sources of allergen cross-contact. The hazard assessment should carefully consider all sources of possible cross-contact, including those related to suppliers and raw ingredients, facility design, product design, allergen sanitation, and employee practices (including errors). Controls for each identified hazard are a key part of the allergen management program. While dedication and separation of manufacturing facilities for allergen-containing foods are ideal, this is often not practical or possible. Operational controls and scheduling can reduce the probability that allergens will end up in products where they should not be.

Allergen sanitation needs to include a plan for allergen removal that has been shown to work (validated) and to include documentation (verification) demonstrating that the sanitation procedure is followed and is done correctly and completely. Visual inspection combined with allergen-specific equipment swabs and finished product testing provides a sensible and robust approach for validation testing.

Periodic and systematic review of all facets of the allergen management program will ensure effective allergen control. Changes occur frequently and rapidly within many food manufacturing environments, requiring the impact of such changes on allergen management to be reassessed regularly. Analysis of sanitation testing results and consumer contact records are examples of data that can be analyzed to inform and refine allergen management programs by early identification of trends that signal problems.

Acknowledgements The author would like to thank Wendy Bedale for assistance in writing this chapter.

References

Branum, A.M., and S.L. Lukacs. 2008. Food allergy among U.S. children: Trends in prevalence and hospitalizations. *NCHS Data Brief* (10): 1–8.

Bucchini, L., A. Guzzon, R. Poms, and H. Senyuva. 2016. Analysis and critical comparison of food allergen recalls from the European Union, USA, Canada, Hong Kong, Australia and New Zealand. *Food Additives and Contaminants Part A - Chemistry, Analysis, Control, Exposure and Risk Assessment.* 33: 760–771.

Deibel, K., T. Trautman, T. DeBoom, W.H. Sveum, G. Dunaif, V.N. Scott, and D.T. Bernard. 1997. A comprehensive approach to reducing the risk of allergens in foods. *Journal of Food Protection.* 60: 436–441.

DunnGalvin, A., C.H. Chan, R. Crevel, K. Grimshaw, R. Poms, S. Schnadt, S.L. Taylor, P. Turner, K.J. Allen, M. Austin, A. Baka, J.L. Baumert, S. Baumgartner, K. Beyer, L. Bucchini, M. Fernandez-Rivas, K. Grinter, G.F. Houben, J. Hourihane, F. Kenna, A.G. Kruizinga, G. Lack, C.B. Madsen, E.N.C. Mills, N.G. Papadopoulos, A. Alldrick, L. Regent, R. Sherlock, J.M. Wal, and G. Roberts. 2015. Precautionary allergen labelling: Perspectives from key stakeholder groups. *Allergy.* 70: 1039–1051.

FARE (Food Allergy Research and Education). 2017. Avoiding cross-contact. Available at: https://www.foodallergy.org/cross-contact#vscontamination. Accessed 7 February 2017.

FARRP (Food Allergy Research and Resource Program). 2017. Components of an effective allergen control plan. Available at: http://farrp.unl.edu/wat/allergen-control-plans. Accessed 3 February 2017.

FSMA (Food Safety Modernization Act). 2011. Public Law 111-353, 111th Congress. Available at: https://www.fda.gov/food/guidanceregulation/fsma/ucm247548.htm. Accessed 23 June 2017.

FSPCA (Food Safety Preventive Controls Alliance). 2016. Preventive controls for human foods participant manual. Available at: http://www.iit.edu/ifsh/alliance/pdfs/FSPCA_PC_Human_Food_Course_Participant_Manual_V1.2_Watermark.pdf. Accessed 30 June 2016.

FoodDrinkEurope. 2013. *Guidance on food allergen management for food manufacturers.* Brussels: FoodDrinkEurope.

Gendel, S.M., N. Khan, and M. Yajnik. 2013. A survey of food allergen control practices in the U.S. food industry. *Journal of Food Protection.* 76: 302–306.

Gendel, S.M., and J.M. Zhu. 2013. Analysis of US Food and Drug Administration food allergen recalls after implementation of the Food Allergen Labeling and Consumer Protection Act. *Journal of Food Protection.* 76: 1933–1938.

GMA (Grocery Manufacturers Association). 2009. *Managing allergens in food processing establishments.* Washington, DC: Grocery Manufacturers Association.

Hewitt, T. 2012. Staying on top of allergen management in your plant. *USDA Small Plant News.* 6: 1–2.

Jackson, L.S., F.M. Al-Taher, M. Moorman, J.W. DeVries, R. Tippett, K.M.J. Swanson, T.J. Fu, R. Salter, G. Dunaif, S. Estes, S. Albillos, and S.M. Gendel. 2008. Cleaning and other control and validation strategies to prevent allergen cross-contact in food-processing operations. *Journal of Food Protection.* 71: 445–458.

Marler, B. 2015. First lawsuit filed over cumin tainted with peanut allergens. Available at: http://www.foodpoisonjournal.com/food-poisoning-information/texas-woman-files-suit-against-reily-foods-over-spice-products-tainted-with-peanut-and-almond-allergens/#.WJinpBsrKUk. Accessed 6 February 2017.

Pozo, V.F., and T.C. Schroeder. 2016. Evaluating the costs of meat and poultry recalls to food firms using stock returns. *Food Policy.* 59: 66–77.

Sicherer, S.H. 2011. Epidemiology of food allergy. *Journal of Allergy and Clinical Immunology.* 127: 594–602.

Taylor, S.L., and S.L. Hefle. 2005. Allergen control. *Food Technology.* 59 (40-43): 75.

Taylor, S.L., S.L. Hefle, C. Bindslev-Jensen, S.A. Bock, A.W. Burks, L. Christie, D.J. Hill, A. Host, J.O. Hourihane, G. Lack, D.D. Metcalfe, D.A. Moneret-Vautrin, P.A. Vadas, F. Rance, D.J. Skrypec, T.A. Trautman, I.M. Yman, and R.S. Zeiger. 2002. Factors affecting the determination of threshold doses for allergenic foods: How much is too much? *Journal of Allergy and Clinical Immunology.* 109: 24–30.

Taylor, S.L., S.L. Hefle, K. Farnum, S.W. Rizk, J. Yeung, M.E. Barnett, F. Busta, F.R. Shank, R. Newsome, S. Davis, and C.M. Bryant. 2006. Analysis and evaluation of food manufacturing practices used to address allergen concerns. *Comprehensive Reviews in Food Science and Food Safety.* 5: 138–157.

USDA (U.S. Department of Agriculture). 2017. The counter-top food safety training program. Available at: https://www.fsis.usda.gov/wps/portal/fsis/topics/inspection/workforce-training/online-references/counter-top-food-safety. Accessed 11 April 2017.

FDA (U.S. Food and Drug Administration). 2004. Food Allergen Labeling and Consumer Protection Act of 2004 (Public Law 108-282, Title II). Available at: https://www.fda.gov/food/guidanceregulation/guidancedocumentsregulatoryinformation/allergens/ucm106187.htm. Accessed 23 June 2017.

———. 2006. Guidance for industry: Questions and answers regarding food allergens, including the Food Allergen Labeling and Consumer Protections Act of 2004. Available at: http://www.fda.gov/food/guidanceregulation/guidancedocumentsregulatoryinformation/allergens/ucm059116.htm. Accessed 7 February 2017.

———. 2015. Current good manufacturing practice, hazard analysis, and risk-based preventive controls for human foods: Final rule. 80 Fed. Reg. 55908. Available at: https://www.federalregister.gov/documents/2015/09/17/2015-21920/current-good-manufacturing-practice-hazard-analysis-and-risk-based-preventive-controls-for-human. Accessed 17 September 2015.

Ward, R., R. Crevel, I. Bell, N. Khandke, C. Ramsay, and S. Paine. 2010. A vision for allergen management best practice in the food industry. *Trends in Food Science & Technology.* 21: 619–625.

Chapter 8
Allergen Cleaning: Best Practices

Lauren S. Jackson

8.1 Introduction

Recent reports have shown an increased prevalence of food allergies, affecting nearly 15 million people in the U.S. (Boyce et al. 2010; Liu et al. 2010). Some individuals with food allergies can develop life-threatening allergic reactions if exposed to minute amounts of allergenic food, and more specifically, the proteins present in allergenic foods. Since there is no cure, strict avoidance of the offending allergen is the only management tool for consumers with food allergies. For this reason, consumers with food allergies depend on accurate and truthful food labels to disclose the presence of allergenic ingredients in the foods they consume. The U.S. and other countries have enacted labeling laws for packaged food in an attempt to protect food-allergic consumers.

The Food Allergen Labeling and Consumer Protection Act (FALCPA) of 2004 requires that manufacturers of all packaged foods sold in the U.S. list major food allergens (milk, egg, peanuts, tree nuts, wheat, soy, fish, and shellfish) when intentionally added to a product as an ingredient (FDA 2016). In 2011, the FDA Food Safety Modernization Act (FSMA) enhanced the FDA's authority to recall foods contaminated with allergens (FDA 2017). FSMA also mandated that FDA issue regulations to require that food facilities implement allergen controls, such as prevention of allergen cross-contact, the unintended introduction of allergens into food. Cross-contact can occur through the transfer of allergens during processing, handling and storage, such as when allergen-containing and non-allergen-containing foods or ingredients are produced in the same facility or processing line and allergen controls are not properly implemented. Causes for cross-contact during manufacture include improper use of product rework, contamination of non-allergenic

L.S. Jackson (✉)

Division of Food Processing Science and Technology, U.S. Food and Drug Administration, Bedford Park, IL 60501, USA

e-mail: lauren.jackson@fda.hhs.gov

© Springer International Publishing AG 2018

T.-J. Fu et al. (eds.), *Food Allergens*, Food Microbiology and Food Safety,
DOI 10.1007/978-3-319-66586-3_8

foods with airborne dust and aerosols generated during production of allergenic foods, reuse of cooking or processing media such as frying oils, and inadequate cleaning of food-contact surfaces (Jackson et al. 2008).

Since the early 1990s, the food industry has devoted considerable resources to developing allergen control plans, with the goal of preventing unintended allergen contamination of food (Taylor and Hefle 2005). Despite these efforts and the enactment of FALCPA and other laws and regulations aimed at ensuring that the food industry improve allergen management during food production, undeclared allergens still remains one of the leading causes of food recalls in the U.S. (Gendel and Zhu 2013; Gendel et al. 2014; USDA, 2016). Food categories that had the highest incidence of recalls include bakery products, confections/candies, dressings, and dairy products (Gendel et al. 2014). An analysis of the root cause of some of the allergen recalls in the U.S. has revealed that the majority of recalls were due to labeling errors, such as placing the food in the wrong package or placing the wrong label on a package of food. In contrast, cross-contact was identified to be the cause of approximately 11% of the recalls for FDA-regulated products (Gendel and Zhu 2013). Although cross-contact is responsible for fewer recalls than labeling associated errors, it is possible that it occurs more often than the recall data might indicate since cross-contact is less likely to be detected than labeling errors. This is due to the fact that cross-contact contamination of foods is likely to be heterogeneously distributed in lots of a food product (Brown and Arrowsmith 2015). It is also important to note that cross-contact is more likely to be responsible for the presence of undeclared allergens in some food categories than others since some foods (e.g., dark chocolate and milk chocolate) are produced on shared processing equipment that are not effectively cleaned to remove allergenic food residues (Bedford et al. 2017).

There have been a number of reports in the literature of consumer complaints stemming from consumption of foods contaminated with undeclared allergens, with cross-contact attributed as a likely cause of allergen contamination. Gern et al. (1991) reported that a number of milk-allergic consumers had adverse reactions after consuming frozen desserts labeled as "nondairy" or "pareve." Analytical testing of the products uncovered that the products contained up to 2200 µg/g milk protein. Although the root cause of the problem was not reported, it is likely that the desserts became contaminated from processing equipment previously used to produce milk-containing products. Similar reports were published by Jones et al. (1992), Laoprasert et al. (1998), Levin et al. (2005), Hefle and Lambrecht (2004), and Spanjersberg et al. (2010) on consumer complaints due to consumption of foods that were contaminated with undeclared allergens due to cross-contact.

Cleaning and sanitation are essential parts of an integrated food safety system that include regulatory compliance, quality standards, hazard analysis critical control points (HACCP), good manufacturing practices (GMPs) and pest control (Cramer 2006). In addition, an effective sanitation program is also critical part of an allergen control plan, particularly in facilities where food products containing allergenic ingredients are manufactured on common lines and shared equipment. Cleaning is only one of a number of practices and procedures for preventing allergen cross-contact (see Chap. 7). However, if it is conducted properly, it can be one of the more effective tools for allergen control. The reader is directed to other publications that discuss the considerations that should be made when developing cleaning programs for the purpose of allergen control (Deibel et al. 1997; Jackson et al. 2008; FARRP, 2008; GMA, 2009; Bagshaw 2009; Brown 2009; FoodDrink Europe 2013). This chapter will discuss the factors that affect removal of allergens from food-contact surfaces, the different approaches for removal of allergenic food soils from food-contact surfaces, best practices and procedures for verifying and validating cleaning treatments for allergen removal, and the steps associated with developing an allergen cleaning program.

8.2 Allergen Cleaning: Basic Concepts

The primary function of cleaning is to remove contaminating soils, prevent biofilm buildup, and prepare food surfaces for sanitizing (Dunsmore et al. 1981; Cramer 2006). Cleaning has traditionally been viewed as a key mechanism for establishing hygiene in a food plant as it used to remove soil and microorganisms and prevent the manufacture of food under unsanitary conditions. More recently, the importance of an effective cleaning program has been identified as one of the best ways to control allergen cross-contact in food production facilities with shared lines or equipment (Jackson et al. 2008). Since cleaning and sanitation are used for various purposes, it is important to understand the difference between cleaning from a microbiological standpoint and what is meant by "allergen clean." The major objective for cleaning from a microbiological standpoint is to render food-contact surfaces free of soil so that they can be effectively sanitized to inactivate microorganisms that may be present. In contrast, cleaning to the point where food-contact surfaces are "allergen clean" means that the allergenic food, and more specifically, the proteins present in allergenic food, are removed. This differentiation is important to point out since surfaces that are microbiologically clean may not necessarily be "allergen clean." Even a small amount of allergen that transfers from an inadequately cleaned surface can result in an allergic reaction in a sensitive consumer. Furthermore, chemical sanitizers and heat which are effective at inactivating pathogens and spoilage microorganisms on equipment surfaces do not cause chemical changes to food allergens that make them safe (Bagshaw 2009; Verhoeckx et al. 2015). Therefore, the only way to clean a surface until it is "allergen clean" is to physically remove the allergenic food soil.

Cleaning from the perspective of removing allergens can be difficult for several reasons. Of the food constituents (proteins, fat, carbohydrate, minerals) that make up food soils, proteins (the constituent most commonly associated with food allergens) are typically the most difficult to remove from food-contact surfaces during cleaning. Proteins in food, particularly when they have been heated and become denatured, tend to adhere to equipment surfaces (Schmidt 2015; Gabrić et al. 2016). Removing allergens from surfaces can also be difficult since some equipment in food facilities was not designed to be cleaned, particularly from the point of being "allergen clean." For example, cloth belting, tunnel ovens, and other equipment were not designed to be wet cleaned since they are either closed systems and/or water and detergents tend to corrode or cause wear to them. Furthermore, use of water to clean equipment used in the manufacture of low water activity foods such as chocolate, nut butters, spices, and other products is often not possible, as the presence of water introduces the risk of microbiological hazards, and reduces product quality. In these situations, surfaces of equipment are often purged with inert ingredients (sugar, salt) or the next food product, or cleaned by other methods (vacuuming, scraping, brush, use of dry steam, grit blasting, etc.), which may not be as effective at removing allergenic residue as use of water (wet cleaning).

8.3 Factors Influencing the Effectiveness of Cleaning Procedures

There are many variables that must be considered when developing an effective allergen cleaning program (Table 8.1). These include the soil characteristics; the composition and type of food-contact surface; how the food was applied to the surface; the type, design, and age of the equipment; and the type of cleaning method. The following subsections provide a more in-depth discussion of soil, equipment, and surface factors. A separate section (Sect. 8.4) provides an in-depth description of the cleaning methods used for removal of allergenic food residue.

8.3.1 Soil Characteristics

Food soils found in the food processing environment can be categorized into four predominant types: carbohydrates, fats, minerals, or proteins (Schmidt 2015). However, food soils are often complex in that they contain more than one of these components. Descriptions of the major classes of food soils, their physical and chemical properties, and most common procedures for removing them are listed in Table 8.2.

Generally speaking, food residues that contain high amount of protein are the most difficult to remove from food-contact surfaces. If the food soil was exposed to excessive amounts of heat, the proteins present will denature and adhere to the surface. Removal of protein soils is typically accomplished with alkaline detergents with and without the presence of oxidizing agents such as sodium hypochlorite or peroxide. The effectiveness of chlorinated alkaline detergents on protein removal is due to their ability to partially hydrolyze and solubilize proteins (Schmidt 2015). Surfactants can be added to these detergents to assist in solubilizing and dispersing soils. In situations where mineral deposits are present in addition to protein soils,

Table 8.1 Variables influencing the effectiveness of allergen cleaning procedures

General variables	Specific variables	Description of variables and their effects on allergen cleaning/removal
Soil characteristics	Physical form (powder, paste, or liquid)	• Pastes can be more difficult to remove than powders and liquids
	Chemical composition (lipid-, protein-, carbohydrate-, or mineral-based)	• Protein-based soils are generally the most difficult to remove, particularly if heat was applied to the soil
	Concentration of allergen in food soil	• Higher concentrations of allergen in the food soil will often necessitate a more intensive cleaning procedure
	Age of soil	• The longer the soil is in contact with a food-contact surface, the more difficult it is to remove
	Processing factors	• Application of heat to the food may result in denaturation of proteins, making them more difficult to remove from some surfaces • Length of processing run causes more soil to build up on equipment surfaces, requiring more extensive cleaning procedures
Food-contact surface	Composition (stainless steel, other metals, plastic, cloth, wood, or glass)	• Cloth surfaces and some metal surfaces can be difficult to clean
	Texture/finish (smooth, textured, or mesh)	• Smooth surfaces are typically easier to clean than those that are rough, textured or have surface defects
Equipment factors	Design of equipment	• Hygienic design of equipment improves cleanability
	Age of equipment	• Surfaces of older equipment may be scratched, scored and have other surface defects that make cleaning more difficult
Cleaning method	Wet cleaning, dry cleaning, or combination of wet and dry cleaning	• Wet cleaning methods that involve the use of detergents tend to be more effective than dry cleaning methods for removing food soils

acidic detergents can be used. However, protein soils should be removed first using alkaline detergents since acid detergents may denature proteins and make them more difficult to remove (Schmidt 2015; Jackson et al. 2008).

In contrast to proteins, carbohydrate-based soils, which are composed of sugars and starch, are fairly water soluble and can be removed from surfaces with water or alkaline-based cleaners (Schmidt 2015). Sugars can caramelize in the presence of heat or react with amino acids to form Maillard products, which can be removed with hot water. Starches present in a high water environment may gelatinize during exposure to heat, making the soil less accessible to detergents (Gabrić et al. 2016).

Table 8.2 Characteristics of food soils and methods for removing them from food-contact surfaces

Food soil	Examples	Characteristics	Heat-induced reactions	Most effective cleaning compounds
Protein-based	Milk protein, egg protein, soy protein	• Can be soluble or insoluble in water • Heat denaturation may increase or decrease solubility in water • Heat denaturation makes some proteins "sticky" and difficult to remove from food-contact surfaces	Denaturation	Alkaline or chlorinated alkaline cleaners
Carbohydrate-based	Starch, sugar, corn syrup	• Typically soluble in water • Component most easily removed during cleaning	Sugars—caramelization or formation of Maillard products with amino acids Starches—interaction with other components	Alkaline cleaners
Fat-based	Oils, fats	• Typically insoluble in water • Easier to remove when cleaning solution is heated to above the melting point of the fat	Polymerization	Alkaline cleaners with and without surfactants/ emulsifiers such as phosphates
Mineral-based	Milk stone, salt	• Polyvalent salts are insoluble in water and alkaline-based cleaning solutions • Heat and hardness of water reduce solubility	Polyvalent salts—interaction with other constituents	Acid-based detergents or cleaning chemicals

Foods that are oil-based may need to be flushed with oil to rinse surfaces before cleaning with water and detergents (Jackson et al. 2008). Fats present in food soils are typically in the form of emulsions, which can be prerinsed from surfaces with hot water above the melting point of the fat (Nikoleiski 2015; Schmidt 2015). More difficult fat and oil residues can be removed with alkaline detergents, which have good emulsifying or saponifying properties.

Cleaning to remove mineral soils depends on the chemical nature of the soil. Calcium and magnesium are often present in mineral films that are very difficult to remove, and in the presence of heat and alkaline pH, they can combine with bicarbonates to form highly insoluble complexes (Schmidt 2015). Other difficult-to-remove mineral deposits contain iron or manganese. Water insoluble mineral films require an acid cleaner for removal, and sequestering agents such as phosphates or chelating agents are often added to hasten film removal.

The length of time that food residues remain on surfaces will result in chemical changes to the soil induced by exposure to heat, UV light, air, enzymes, and microorganisms; all of these changes impact soil removal (Wilson 2005; Gabrić et al. 2016). As the soil ages, it becomes more difficult for cleaning detergents to penetrate the soil. Furthermore, the risk of biofilm formation also increases with the length of time food soils remain on surfaces. Biofilms can be difficult to remove and usually require cleaners as well as sanitizers with strong oxidizing properties (Schmidt 2015).

The physical form of the allergenic food soil (solid, liquid, paste, or particulate) will also be an important aspect that should be considered when designing an allergen cleaning procedure. For example, a different method may be needed for food pastes such as almond butter than defatted almond flour; food pastes tend to adhere to food-contact surfaces, making them difficult to remove. Although some dry food powders are relatively easy to remove, they can create problems since they are likely to disperse in the air and settle on pieces of equipment throughout the production area (Moerman and Mager 2016). Particulate food materials (i.e., whole nuts, nut fragments, fragments of cereal bars containing allergens, etc.) are a particular challenge from a cleaning standpoint since they can become trapped in some equipment. This is of concern since these particulates may contain enough allergenic material to cause an allergic reaction in a food-allergic consumer if they become dislodged and contaminate the next food produced on the processing line (Brown and Arrowsmith 2015).

8.3.2 Food-Contact Surface and Equipment Factors

Food processing operations that rely on wet cleaning procedures to ensure product quality and safety require that equipment used be of hygienic design (Nikoleiski 2015). In contrast, requirements to control microbial and allergen hazards through wet cleaning are absent in low-moisture food operations. Consequently, the equipment in these facilities is not usually designed for wet cleaning, and the hygienic standards applied are lower than those for processors that use wet cleaning (Nikoleiski 2015). Excellent descriptions of hygienic design of equipment used in food processing facilities (i.e., heating equipment, air-blast freezing systems, pumps, valves, etc.) can be found in Lelieveld et al. (2016).

All equipment used in food production should be designed to permit easy and rapid access to the interior for cleaning and sanitation and should be self-draining to ensure that residues from food or cleaning operations can be discharged (Stone and Yeung 2010; Nikoleiski 2015). For equipment that needs to be disassembled before cleaning, dismantling should be made easy so that hidden areas of the equipment can be adequately accessed and cleaned. Older equipment (i.e., built prior to 1960) was not designed with allergen cleaning in mind, so modifications should be made to enable visual inspection of surfaces for presence of food residue. Permanent joints, such as welds, are preferable to dismountable joints, and should be smooth, continuous, and free of overlaps (Nikoleiski 2015).

From a sanitary perspective, food-contact surfaces should be smooth, impervious, free of cracks and crevices, nonporous, nonabsorbent, noncontaminating, nonreactive, corrosion resistant, durable and easy to clean (Dunsmore et al. 1981; Schmidt et al. 2010). Stainless steel is the predominant material of choice for food processing equipment since it is smooth, durable, is corrosion resistant and can be easily cleaned (Schmidt et al. 2010; Gabrić et al. 2016). However, perforated or mesh stainless steel belting or welding points of stainless steel can harbor food soils and be more difficult to clean. Other types of surfaces present in food production facilities include plastics (PVC, polycarbonate, urethane, UHMW, vinyl), rubber, wood and cloth or fabric. Plastic surfaces typically are easily cleaned, particularly when they are new and in good condition. However after time they wear and become damaged, dramatically increasing the chance for cross-contact (Gabrić et al. 2016). Wood and cloth surfaces are to be avoided, since they are highly porous and difficult to clean. The finish and smoothness of the food-contact surface and the surface condition (e.g., pitted, cracked, or scratched) also can vary. In general, surfaces that are smooth and nonporous are more readily cleanable. Surfaces of older equipment tend to have more scratches, pitting, and other defects which tend to harbor food residue and prevent thorough cleaning.

8.4   Cleaning Methods

The three major categories of cleaning procedures used to remove allergenic food soils include wet cleaning, dry cleaning, and product purging or push-through. The most powerful tool for removing allergens from food processing equipment is water since water-based detergent solutions tend to be very effective at dissolving and removing proteins from food-contact surfaces. In most cases, the ability to use water is a function of the water activity of the food being produced at that process point. For example, water is used to clean at process points where product is in the form of wet mixes, but is used less frequently for cleaning at points where the food has low water activity (e.g., milk powder after spray drying, baked goods such as crackers or cookies after baking) (Jackson et al. 2008). Although traditional water-based cleaning and sanitizing methods are very effective at reducing the presence of microorganisms in most processing environments, they are not recommended for use in the low-moisture food processing areas since introduction of water may be conducive for survival and growth of pathogenic microorganisms (Beuchat et al. 2013). Facilities that process high-water-activity foods are designed to accommodate water, have drains to allow for drainage of water after cleaning, have equipment that can be disassembled, and electronics wired to withstand moisture. Floors and walls in wet facilities have smooth surfaces to prevent accumulation of allergens, prevent microbial growth, and allow for cleaning (Jackson et al. 2008).

8.4.1 Wet Cleaning

Wet cleaning refers to the use of water alone, or in combination with detergents and other cleaning chemicals, to remove food soils (Schmidt 2015). The major types of detergents and cleaning chemicals include the following: alkaline detergents, alkaline detergents containing oxidizing agents, acid detergents, enzyme-based cleaners, and sanitizers. Surfactants are typically added to detergents to assist in cleaning by wetting and then dispersing the soil in the cleaning solution (Schmidt 2015).

As mentioned previously, food soils vary in composition and manner in which they are applied to food-contact surfaces; thus, wet-cleaning protocols should be designed for each soil and piece of equipment. Chlorinated alkaline detergents are the most commonly used and effective detergent for removal of protein soils from food-contact surfaces. Other detergents can be used if they are first evaluated for effectiveness for soil removal. Disinfection/sanitization of food-contact surfaces is typically accomplished with a number of chemical sanitizers (chlorine-based, iodophors, quaternary ammonium compounds, organic acids, anionic sanitizers, peroxy acid compounds). Descriptions of the classes of cleaning chemicals/solutions and their ability to remove protein soils are listed in Table 8.3. The key factors for effective wet cleaning include the length of *time* of the cleaning step or cycle, *action* or mechanical force, *chemical* composition and concentration, and *temperature* (TACT). A description of these factors and a summary of their effects on wet cleaning procedures are found in Table 8.4.

Wet-cleaning methods can be divided into four categories: clean-in-place (CIP), where equipment requires minimal or no disassembly and the cleaning and sanitizing solutions are recirculated throughout the line; clean-out-of-place (COP), where equipment can be disassembled and the loose parts are cleaned and sanitized in tanks; foam or gel cleaning, where the detergent in the form of gel or foam is sprayed on equipment surfaces; and manual cleaning (Bagshaw 2009). The choice of the wet cleaning method and frequency with which it is used will depend on the type of operation and the types of soils involved.

Table 8.3 Types of cleaning chemicals/solutions and their effectiveness at removing protein soils

Type of cleaning solution	Examples	Effectiveness at removing protein soils	Comments
Alkaline + oxidizer	Sodium hydroxide in combination with sodium hypochlorite or peroxide	Excellent	These types of detergents can partially hydrolyze and solubilize proteins.
Alkaline	Sodium hydroxide or potassium hydroxide	Fair to very good	Alkaline detergents are effective at cleaning fats and oils from food-contact surfaces. They are also fairly effective at cleaning some of the more water soluble proteins.
Acid	**Inorganic acids**: phosphoric, nitric, sulfamic, sodium acid sulfate, and hydrochloric. **Organic acids**: hydroxyacetic, citric, and gluconic	Poor	Acid detergents may precipitate proteins on surfaces; they typically are used to remove mineral soils after first removing proteins with alkaline detergents
Enzyme-based	Amylases and other carbohydrate-degrading enzymes, proteases, and lipases	Excellent for proteases	The length of time needed for enzymes to be effective limits the use of enzymes; contact time for removing protein deposits may take minutes to several hours.
Sanitizers	Chlorine-based (e.g., sodium hypochlorite), iodophors, quaternary ammonium compounds, organic acids, anionic sanitizers, peroxy acid compounds	Poor–good	Sanitizer solutions may be effective at removing loosely associated protein soils, but will not be effective at removing insoluble and precipitated protein soils.
Water	–	Poor–good	Water may be effective at removing loosely associated protein soils, but will not be effective at removing insoluble and precipitated protein soils.

Table 8.4 Factors influencing the effectiveness of wet cleaning procedures

Factor	Considerations
Time	• Too little time results in not enough interaction with soil and thus, inadequate soil removal • Too much time can result in a cleaning solution to cool to a temperature where it is not effective at removing soil • Proper time exposure results in wetting of soil and removal from surfaces
Action	• Can include the use of manual force and automated cleaning systems • Loosens soils and disrupts biofilms • Most effective when all surfaces are impacted
Chemical (composition and concentration)	• Improper choice of cleaning chemical may result in inadequate removal of soil • Concentration of chemicals should be checked with sufficient frequency to ensure that they are at desired strength • Use of cleaning solutions that are too dilute will result in inadequate removal of soil • Use of cleaning solutions that are too concentrated will waste money, may result in detergent deposition on surfaces, and may corrode equipment surfaces • Use of cleaning solutions at proper concentrations will result in removal of soils
Temperature	• Cleaning solution that is below optimal temperature may not dissolve the food soil • Use of cleaning solutions that are too hot may result in precipitation of the protein on the food-contact surface • Proper cleaning solution temperature [as indicated in the sanitation standard operating procedure (SSOP)] results in adequate dissolution and removal of soil

Adapted from PennState Extension (2017)

The method of choice for cleaning large tanks, kettles, or piping systems which cannot be accessed for manual cleaning is CIP. The process involves circulation of detergent through equipment by use of a spray ball or spray nozzle to create turbulence (Cramer 2006). The flow rate, concentration of cleaning chemicals, volume of cleaning fluids, and time are predetermined through validation studies. CIP procedures tend to be very effective once validated since they are fully automated and can be applied consistently (Jackson et al. 2008; Stone et al. 2009). Reuse of some of the cleaning solutions used in CIP, particularly the detergents and final rinse water, is a common practice (Bagshaw 2009; Valigra 2010). As reused detergents may have excessive buildup of proteins, their cleaning effectiveness may be compromised (Merin et al. 2002; Bagshaw 2009; Du et al. 2011). Similarly, allergenic food proteins present in the reused detergent may carry over to other equipment during cleaning.

COP procedures involve cleaning disassembled equipment parts in a COP tank, which contains the cleaning solution (detergent, sanitizer) (Cramer 2006; Bagshaw 2009). COP tanks are designed to circulate cleaning solution and provide agitation and temperature control during cleaning procedures. For COP procedures to be effective, it is essential that all parts are completely submerged to ensure adequate exposure to the cleaning chemicals. Similar to CIP, COP procedures can be automated and are very effective at removing food soils once they are validated (Cramer 2006; Bagshaw 2009).

Foam cleaning is a common method for cleaning equipment and rooms (Bagshaw 2009). Detergent foams or gels are used for cleaning of equipment, when contact time is important (Nikoleiski 2015). For these cleaning procedures, foams and gels of concentrated detergents are applied to equipment surfaces, and allowed to remain in contact with the soil and surfaces for a period of time that enables difficult-to-clean soils to be solubilized. The surfaces are then rinsed with water to remove detergent and soil. Use of high pressure hoses to rinse equipment is discouraged since they can spread allergens throughout the facility. Advantages of foam cleaning over manual cleaning include (1) ability to apply detergent solution to large and less accessible areas, (2) longer contact time between detergent and soil, (3) reduction in the time and manpower needed for cleaning, (4) control of detergent used, and (5) safer application of potentially hazardous detergents (Bagshaw 2009).

Manual wet cleaning of equipment and surfaces is the most common method utilized by the food manufacturing industry due to its flexibility and ability to spot-clean difficult-to-clean areas (Bagshaw 2009). It is accomplished by applying detergent to the food-contact surface and manually scrubbing the surface with clothes, brushes, and scrapers to remove soils. As with other cleaning methods, the choice of cleaning chemical (type, concentration, and temperature) is critical to removal of soils. Brushes, scrapers, and other equipment used for manual cleaning should be dedicated, if possible, to minimize the risk of cross-contact. Smith and Holah (2016) wrote an in-depth review on the selection, use, and maintenance of equipment used in manual cleaning operations.

8.4.2 Dry Cleaning

Dry cleaning, the removal of food soils without water or water-based cleaning chemicals, is used in food production facilities when the presence of water could compromise the quality and consistency of the product or create conditions that enhance microbial growth (Jackson et al. 2008; Stone et al. 2009; Stone and Yeung 2010; Moerman and Mager 2016). Wet cleaning for equipment present in low moisture areas of a facility may be used in occasion, but only when equipment can be disassembled, and the wet cleaning operations are performed in another location, where presence of water can be tolerated. Dry cleaning is also more widespread in older food production facilities that were not originally designed based on current hygienic design principles (Moerman and Mager 2016). Examples of product lines where dry cleaning procedures are used to control microbiological and allergen hazards include chocolate, baked goods, cereals, and powdered beverages. Methods that are used to remove food residue in dry facilities include brushing, scraping, wiping, use of pigging systems (described below), dry steam cleaning, and use of vacuums (Stone and Yeung 2010; Nikoleiski 2015; Moerman and Mager 2016). Use of compressed air can be effective at dislodging particulate materials from hard to reach areas. However, the use of air hoses should be avoided because it may disperse allergenic materials into the surrounding environment (Jackson et al. 2008; Bagshaw 2009).

Dry cleaning procedures vary in effectiveness depending on the type of food soil, type of equipment and food-contact surface, and specific cleaning method. Loosely attached materials and dry powders tend to be easier to remove than cooked on materials and food pastes (Jackson et al. 2008). High-fat foods and food pastes are notoriously difficult to remove solely with dry cleaning methods such as scraping and brushing. In these cases, dry cleaning is often combined with other methods such as wiping with moistened cloths (for accessible areas), product purging, and use of pigging systems.

Manual scrapers and brushes are frequently used in dry food manufacturing facilities since they can be used to spot-clean equipment and can be very effective at removing strongly adhered and loosely associated food soils, respectively. Brushes and scrapers should be dedicated for use with allergenic food residue and color coded to ensure that they are not used in allergen-free locations in the facility. If not used properly, brushes and scrapers can generate dusts of allergenic food soils which can recontaminate equipment (Moerman and Mager 2016). Since brushing and scraping only lifts soils from surfaces, vacuums are often used in combination with these manual cleaning techniques to remove the food residues. Single-use wipes prewetted with sanitizer or alcohol can be used to remove remaining food residue on equipment surfaces after brushing or scraping.

In facilities with piping and other transport lines, cleaning can be conducted with pigging systems. Pigs, devices made of food-grade materials such as silicone or natural rubber, are often used to push out dry food residues present in long expanses of piping (Moerman and Mager 2016). Pigs typically have a diameter that is slightly larger than the inner diameter of the piping. As a result, inner surfaces of piping are scraped as pigs are propelled through the system with compressed air or inert gas. Although pigging can be very useful for removing residual dry food powders from piping, the technique alone is typically not successful at controlling allergens since pigs may occasionally slip over the product and leave residue behind (Moerman and Mager 2016). Product or ingredient flushing or purging is often used after pigging to remove the remaining food residue.

Vacuuming is a very effective tool for removing dry food soils, dust, and dirt from floors, walls, and equipment. High-efficiency particulate air filtration vacuum systems (central and portable) with HEPA filters capable of removing particles of 0.3 μm at 99.97% efficiency have been developed to remove and contain dust and debris during dry cleaning of food plant areas (Jackson et al. 2008). Vacuum systems are very effective at removing visible food particles, but they are not effective at in-depth cleaning or removing dried or baked-on soils unless food-contact surfaces are first scraped or brushed (Jackson and Al-Taher 2010; Moerman and Mager 2016). Similar to other tools and equipment used in dry cleaning, vacuums should be dedicated to one area and one type of application to prevent transfer of allergens to other locations in a shared facility.

Disposable (single-use) cloth or paper wipes saturated with water, alcohol or sanitizer solutions (i.e., quaternary ammonium compounds) have been used to clean food-contact surfaces in areas that have been first vacuumed to remove the bulk of the food soil. The advantage of these moistened wipes is that they localize water and minimize dust generation (Jackson et al. 2008). In laboratory studies, Jackson and Al-Taher (2010) found that alcohol-saturated, single-use cloths were very effective at removing a variety of allergenic food soils (nonfat dry milk; soy flour, soy-based infant formula, peanut flour, whole egg) when baked onto the surface of stainless steel, urethane, and Teflon coupons.

A current trend in the food industry is to develop and use chemical-free cleaning methods to reduce costs, protect workers and prevent chemical carryover (Powitz 2014). One approach to chemical-free cleaning is use of "dry" steam to remove soils from and sanitize food-contact surfaces. Commercial dry steam-vacuum systems have been developed for automatic cleaning of flat conveyor belting. Dry steam is produced by a jet of superheated, vaporized water, which impinges on the surface of the belt and loosens food soils, while the vacuum manifold removes the debris and dries the surface of the belt (Moerman and Mager 2016). A limited study on the use of a dry steam-vacuum system found that its effectiveness depended on the nature of the food soil; dried egg soil was particularly difficult to clean from urethane-faced belting using the dry steam-vacuum treatment (Al-Taher et al. 2011).

In summary, although many techniques are available for cleaning equipment in dry processing facilities, it is often difficult to dry clean sufficiently to the point where surfaces are "allergen clean." Research is needed to evaluate current dry cleaning methods and identify conditions and techniques that could be used to render surfaces that are free of allergenic food residue. Research should also be directed at developing new nonaqueous detergents and other methods for allergen cleaning in the dry processing environment.

8.4.3 Purging or Push-Through Cleaning

Product or ingredient purge or "push-through" cleaning is running the next solid, semisolid, or liquid product, or an ingredient through a line in an effort to remove the residue left from prior production (Stone and Yeung 2009). This cleaning procedure is used most frequently when the product is a food paste containing a high fat content, for removal of caked-on food residue, and when surfaces that need to be cleaned are enclosed (e.g., pipes) and not accessible (Moerman and Mager 2016). Some examples of product lines that are purged using push-through cleaning include chocolate and nut butters. Ingredients that are often used to purge lines that cannot be cleaned with other more conventional methods include salt, sugar, granular starch, and flour. Oil flushes are occasionally employed to flush piping and other equipment that previously conveyed other foods of high fat content. Validation studies are needed to define the conditions (volume or weight, flow rate, and temperature of purge material) needed to remove allergenic soils during purging treatments (Stone et al. 2009). Reports by Taylor and Hefle (2005) and Zhang et al. (2013) illustrate the difficulty in use of product push-through to remove highly viscous, sticky foods from processing equipment. For these situations, dedicated equipment may be the only way to realistically reduce the risk of allergen cross-contact (Stone et al. 2009).

Dry ice (solid CO_2) pellets, soda (sodium bicarbonate) blasters, and grit blasting are techniques used for removal of baked-on residues and caked-on dust from food-contact surfaces in low water environments (Moerman and Mager 2016). Blasting techniques are very effective for hard, nonelastic soils, but are less effective when food residues are soft and elastic. The advantage of these methods is that they can be used to clean and remove most soils without damaging delicate surfaces. However, blasting techniques can disperse soils and do not capture the soil removed from the surface (Jackson et al. 2008).

8.5 Cleaning Validation and Verification

Cleaning validation, for the purpose of allergen removal, refers to the process of collecting and evaluating technical information or data to ensure that that a defined cleaning procedure is able to effectively and reproducibly remove the allergenic food from the specific food processing line or equipment, or reduce the amount of allergens to an acceptable level (Jackson et al. 2008; Stone et al. 2009; Cochrane and Skrypec 2014). Validation studies are useful in that they not only provide information on the effectiveness of cleaning procedures, they also identify areas and pieces of equipment which are difficult to clean and which need other approaches for ensuring allergens are removed (Jackson et al. 2008).

Ideally, cleaning procedures should be developed and validated before commercial manufacture of a product begins and any time changes are made to the manufacturing (longer processing run; change in ingredients; change in processing temperature; scheduling changes, etc.) or cleaning procedures (changing detergent type, concentration, temperature, etc.) which might influence cleaning efficacy (Jackson et al. 2008; Stone and Yeung 2009). Although identical products may be produced on different production lines, each line should be assessed independently if there are design differences which may impact cleanability (Nikoleiski 2015). As with most validation procedures, the best approach is to revalidate cleaning procedures at least once per year or at another prescribed amount of time (Lopez and Morales 2015). Yearly reassessments are needed due to subtle changes in ingredients/product, processing conditions, and equipment design or setup that might influence the effectiveness of cleaning procedures. Revalidation may also be warranted when product labeling changes are made to the product and when there are technical advances in analytical methods used for validating cleaning procedures.

Validation involves physical (visual) inspection of all accessible food-contact surfaces, as well as analyzing swab samples of equipment surfaces and samples of final
rinse water during a CIP cleaning procedure, push-through or purge material, and/or
the product produced after the changeover for the presence of allergenic food residue
(Jackson et al. 2008; Stone et al. 2009; Brown 2009; Cochrane and Skrypec 2014).

Cleaning verification refers to the process of demonstrating that validated cleaning protocols are properly implemented and that the procedures are working as
planned once the commercial manufacture of a product begins (Jackson et al. 2008;
Stone and Yeung 2009). Verification, which can be accomplished at the conclusion
of each change-over or at a predefined schedule, provides assurances that the line is
"allergen clean." Verification activities include many of the same practices and procedures (visual inspection, use analytical tests such as protein swabs, adenosine
triphosphate swabs, allergen-specific lateral flow devices, etc.) used for validation
(Jackson et al. 2008; Stone and Yeung 2009).

8.5.1   Visual Inspections for Validation and Verification of Cleaning Procedures

Validation and verification of cleaning treatments typically begins with a visual
inspection of equipment to ensure that all surfaces are visibly clean. Areas that
should be inspected include those most difficult to clean, such as equipment surfaces where foods are heated and locations which may harbor allergenic food residue (corners, o-rings or gaskets, and equipment that may have surface defects such
as scratches or tears) (Jackson et al. 2008; Stone and Yeung 2009). Zones above the
processing area, walls, and floors should also be inspected for presence of residual
soils. Visual detection of food residue indicates failure of the cleaning procedure
and that additional cleaning is needed. Although visual inspection is a valuable tool
in establishing whether procedures are effective, this practice has limitations. It is
often not possible to visually inspect all food-contact surfaces since some areas are
not accessible. Inadequate lighting and the colors and textures of some surfaces
make soils difficult to detect (Jackson et al. 2008).

8.5.2   Obtaining Samples for Validation and Verification of Cleaning Procedures

Statistically based sampling plans should be developed for obtaining samples used
to evaluate the effectiveness of cleaning procedures. The purpose of the sampling
plan is to maximize the probability of detecting allergenic food if it is present on
food-contact surfaces, in push-through material, and in the next product manufactured on the production line (Brown and Arrowsmith 2015). Factors that need to be
considered in developing a sampling plan include when, where and how the samples

are obtained, as all of these parameters can influence the results of a validation study (Brown and Arrowsmith 2015). Another factor that should be considered is the nature of the allergen (particulate, powder, paste, etc.). Outcomes of a poor sampling plan are that allergens are not detected in samples and inadequate cleaning procedures are approved.

Swab samples should be taken from difficult-to-clean areas (seams, areas where there might be burn-on of food, valves, etc.) as well as other areas on each piece of equipment. If multiple lines are used, it is important to obtain swab samples from equipment in all lines. It is also critical to obtain swabs of equipment prior to cleaning to serve as positive control samples, and to ensure that swabs used to sample equipment surfaces are appropriate for use and that procedures are in place to prevent contamination with allergens prior to and after sampling (Sheehan et al. 2012; Baumert and Taylor 2013; Baumert 2014).

Final rinse water obtained at the end of a CIP cleaning treatment can be sampled when there are no ways to access equipment surfaces, for example, in closed or piping systems (Brown 2009). However, testing rinse water for the presence of allergenic food residue may be difficult due to the presence of residual cleaning solutions (detergents, sanitizers) which can affect analytical results. Furthermore, allergens present in rinse water may be diluted to the point where they are present at concentrations below the detection limit of the analytical test (Brown 2009).

In dry systems, where lines are purged with the next product or inert materials (salt, sugar, etc.), a sampling plan is needed to determine where, when, and how much of each sample is obtained to ensure that allergenic residues are removed from the system. Samples of the push-through material should be obtained at multiple locations in the processing line (at different pieces of equipment) and after different volumes of material are used to purge the system. If multiple lines are used, it is important to obtain push-through samples for all processing lines. Obtaining samples of the first product produced on a manufacturing line after a changeover is essential for cleaning validation studies since any residue that remained in the line will likely be present in this product (Brown 2009). Samples should also be obtained in product produced in the middle and end of the processing run to ensure that they are free of allergenic food residue.

Research is needed to develop statistically valid sampling plans that could be used for obtaining swab, push-through, and final product samples for allergen testing. Having such sampling plans will be useful in ensuring that cleaning procedures are effective. This research is even more essential if a risk-based approach for labeling foods for allergens is to be considered in the future (Brown and Arrowsmith 2015).

8.5.3 Analytical Methods Used for Validating and Verifying Cleaning Procedures

During validation studies, it is important to accompany visual inspections with analytical tests as a way to confirm that surfaces that are "visually clean" are actually "allergen clean." Analytical methods that are available fall into two main categories:

nonspecific tests and specific tests. Nonspecific methods are those that do not actually target the allergenic food or allergenic proteins, but rather assess the presence of surrogate compounds. At present, the most frequently used nonspecific analyses are adenosine triphosphate (ATP) bioluminescence and total protein swabs or tests. ATP swabs, typically used as general hygiene indicators, are available from a number of vendors for detecting the presence of ATP from food/food residues and microorganisms remaining on equipment after cleaning/sanitization steps. With the use of a handheld luminometer, ATP swab results can be obtained on-site and in <1 min. In comparison to allergen-specific methods, ATP swabs are relatively inexpensive. Since ATP swabs do not actually detect the allergenic food or protein, the validity of this method for allergen cleaning validation is questionable. Jackson and Al-Taher (2010) found no correlation between ATP test results and allergen-specific LFD tests when used for determining the adequacy of dry cleaning methods. Therefore, the method is used mainly as a way to obtain a general gauge on the adequacy of cleaning procedures.

Similar to ATP tests, total protein swabs are used mainly as general hygiene indicators since they indicate the presence of protein on food-contact surfaces and in final rinse water from all possible sources: microbiological, food, dust, etc. Protein swabs are relatively inexpensive compared to the more specific methods (immunochemical, DNA-based, mass spectrometry), are fairly sensitive in that some swabs are designed to detect <10 µg of protein on surfaces or in solution, and have a turnaround time of <10 min. They also are useful when more specific immunochemical assays are unable to detect allergenic proteins due to conformational changes that occur during thermal processing (Brown 2009). Jackson and Al-Taher (2009) compared allergen-specific (ELISA) and nonspecific (ATP, total protein) methods for detecting the presence of soy-based food residues in solution and on food-contact surfaces. They found that a total protein swab had the lowest detection limits for soy flour, soy milk and soy-based infant formula dried onto the surface of stainless steel coupons (80 °C, 1 h) compared to ATP and soy-specific ELISA methods. Although in this specific application total protein swabs were more sensitive than ELISA, this may not be the case for other allergenic food residues. Therefore, it is important to evaluate whether protein swabs as well as other analytical tests can detect each allergenic food prior to cleaning validation studies.

The most specific and most powerful analytical method for determining the presence of allergenic food on surfaces or in water or food samples are immunochemical assays that target either the allergenic proteins themselves or other protein(s) in the sample. Formats of immunochemical assays currently available for testing samples obtained during cleaning validations include enzyme-linked immunoassays (ELISA) and lateral flow devices (LFDs). ELISA kits and LFDs, which can be purchased from a variety of manufacturers, can be used to detect most of the major allergens with the exception of some tree nuts and some fin fish species. Quantitative ELISA kits which can be used to quantify allergens in finished product, rinse water, and push-through materials have limits of detection in the low ppm (µg/g) range (Jackson et al. 2008). Although qualitative well ELISA assays are available for testing swab samples and rinse water, they have largely been replaced by the use of LFD tests. LFD tests are sensitive, relatively inexpensive, easy to use, and provide results in <5 min. The detection of allergens in the finished product or push-through materials by quantitative ELISA or swab and rinse water by LFD tests generally indicates failure in the design and execution of the cleaning procedures (Jackson et al. 2008).

Although ELISA and LFD tests are reliable tools for detecting allergens in foods and samples obtained in cleaning validations, they have limitations. Detection of allergens is achieved by binding of the target proteins in the sample with antibodies associated with the ELISA and LFD tests. Any changes in the structure or conformation of proteins can dramatically reduce their solubility and ability to bind to these antibodies, thereby affecting test results (Jackson et al. 2009). Thermal processing, fermentation, hydrolysis, changes in pH, and presence of strong oxidizing agents (present in chemical sanitizers and detergents) have all been shown to affect the ability to detect proteins with immunochemical methods (van Hengel 2007; Jackson et al. 2008; Fu et al. 2010; Taylor et al. 2009; Yeung 2009; Fu and Maks 2013). High fat food matrices, presence of interfering compounds (i.e., polyphenolic compounds), and cross-reactivity with structurally similar proteins are other factors that influence test results. Furthermore, analytical targets differ between ELISA kits, and the lack of standard reference materials makes comparison of results obtained with differ kits difficult (Stone and Yeung 2009). One must be mindful of the limitations and evaluate the ELISA and LFD tests of choice for their ability to detect allergenic food residues before starting cleaning validation studies.

Most commercial immunochemical methods available for the detection of food allergens are analyte-specific in that only one allergen can be detected in an analysis. As a result, several kits are needed when testing foods or food-contact surfaces for multiple allergens that may be present. To reduce the time and cost associated with use of multiple kits, multi-analyte immunochemical methods are being developed. Currently, one LFD is commercially available for the detection of multiple tree nuts in a single assay (Neogen 2017). A limitation of this assay is that it does not indicate which of the tree nuts is present when a positive response is registered. The multi-analyte profiling (xMAP®) technology, a commercial multiplex test kit based on the use of established antibodies, was recently developed for the simultaneous detection of up to 14 different food allergens and gluten (Cho et al. 2015). The assay simultaneously detects crustacean seafood, egg, gluten, milk, peanut, soy, and nine tree nuts (almond, Brazil nut, cashew, coconut, hazelnut, macadamia, pine nut, pistachio, and walnut) (Cho et al. 2015). Further work is needed to validate this system and determine its applicability in detecting allergens on swab samples and other samples used in validation of cleaning procedures.

Polymerase chain reaction (PCR) kits are occasionally used to assist in validating cleaning methods, particularly when ELISA kits are not available for a particular allergenic food (some tree nuts; fin fish) (Holzhauser and Röder 2015). PCR methods are specific in that they detect the presence of DNA from the allergenic food, and they can be multiplexed so that multiple allergenic foods can be detected in one assay. These assays have several advantages over immunochemical methods such as ELISA in that DNA has greater stability than proteins during food processing and matrix effects have less of an impact on PCR methods (Baumert 2014). However, a major drawback of PCR is that absence of DNA does not indicate whether surfaces are "allergen clean" with respect to the presence of allergenic proteins (Brown 2009).

A relatively new technology for detecting allergens is mass spectrometry (MS), a method that can identify proteins and peptides with a high level of accuracy and sensitivity. Unlike other methods, MS directly detects allergenic proteins and through its inherent high resolving power, can differentiate between closely related proteins and peptides. Unlike immunochemical methods such as ELISA, conformational changes that occur in proteins due to thermal processing are not likely to affect detection using MS. MS methods are inherently multiplexed since they can be used to detect multiple allergens in a single analysis (Monaci and Visconti 2009). Although there have been numerous publications on the development of MS methods for detecting allergens in food (Chassaigne et al. 2007; Monaci et al. 2014; Parker et al. 2015), the use of MS is limited at this time due to the high cost of MS instruments, the length of time needed to prepare samples for analysis, the high level of expertise needed to perform an analysis, and the difficulty in using the method to quantify allergens in food. It is anticipated that MS methods may be used to assist in allergen cleaning validation studies when some of these obstacles are overcome.

8.6 Development of an Allergen Cleaning Program

Once the steps of the cleaning process and validation and verification procedures are identified and optimized, the entire process should be fully documented and incorporated into an allergen cleaning program (Jackson et al. 2008; Stone and Yeung 2009). This program consists of three parts: sanitation standard operating procedures (SSOPs), validation procedures, and verification procedures. The SSOPs should be as detailed as possible and should include (1) a description of the range of application for the SSOPs, equipment, and products, (2) identification of who is responsible for performing the cleaning operations, (3) a detailed description of the cleaning procedure(s), and (4) records kept for each cleaning procedure. The validation/verification procedures should (1) define the intention and scope of validation/verification, (2) describe the analytical procedures to be used for validation and verification, (3) define the sampling procedures and the reasons for using them, (4) define the final acceptance criteria for cleaning validation/verification, (5) describe corrective actions when the cleaning procedures do not meet the acceptance criteria, and (6) describe records kept for validation/verification procedures. The overall cleaning program should be evaluated periodically to reassess its effectiveness.

8.7 Summary and Conclusions

To ensure that foods are safe for food-allergic consumers, controls must be in place to prevent allergen cross-contact during storage, transport, and production of food. Although cleaning is one of many allergen control strategies, it is one of the more powerful ones at preventing cross-contact if cleaning procedures are properly performed. Development and implementation of effective cleaning procedures requires one to understand the many factors that affect removal of allergenic food residues from the food production environment. Although wet cleaning methods tend to be effective at removing allergens from food-contact surfaces, all procedures must be thoroughly evaluated to ensure that they can achieve the desired effect. In contrast, allergen cleaning in dry food environments can be challenging since allergenic food soils are difficult to remove in the absence of water and detergents. Research is needed to develop new approaches for cleaning in a dry foods environment that remove allergenic food residue yet do not introduce microbiological hazards. Although there are a number of analytical tools currently available for detecting allergens in food and on food-contact surfaces, research is needed to develop sensitive, inexpensive, and rapid multiple-allergen detection methods that could be used for validating and verifying allergen cleaning procedures.

References

Al-Taher, F., C. Pardo, and L. Jackson. 2011. Use of a dry steam belt washer for removal of allergenic food residue. International Association for Food Protection (IAFP) Annual Meeting, July 31, 2011–August 3, 2011, Milwaukee, WI.

Bagshaw, S. 2009. Choices for cleaning and cross-contact. In *Management of food allergens*, ed. J. Coutts and R. Fiedler, 114–137. Oxford: Wiley-Blackwell.

Baumert, J., and S. L. Taylor. 2013. Best practices with allergen swabbing. *Food Safety Magazine*. June/July edition. Available at: http://www.foodsafetymagazine.com/magazine-archive1/june-july-2013/best-practices-with-allergen-swabbing/. Accessed 17 April 2017.

Baumert, J. 2014. Detecting and measuring allergens in food. In *Risk management for food allergy*, ed. C.B. Madsen, R.W.R. Crevel, C. Mills, and S.L. Taylor, 215–226. Oxford: Elsevier.

Bedford, B., Y. Yu, X. Wang, E.A.E. Garber, and L.S. Jackson. 2017. A limited survey of dark chocolate bars obtained in the United States for undeclared milk and peanut allergens. *Journal of Food Protection*. 80: 692–702.

Beuchat, L.R., E. Komitopoulou, H. Beckers, R.P. Betts, F. Bourdichon, S. Fanning, H.M. Joosten, and B.H. Ter Kuile. 2013. Low-water activity foods: Increased concern as vehicles of foodborne pathogens. *Journal of Food Protection*. 76: 150–172.

Boyce, J.A., A. Assa'ad, W. Burks, S.M. Jones, H.A. Sampson, R.A. Wood, M. Plaut, S.F. Cooper, M.J. Fenton, S.H. Arshad, S.L. Bahna, L.A. Beck, C. Byrd-Bredbenner, C.A. Camargo Jr., L. Eichenfield, G.T. Furuta, J.M. Hanifin, C. Jones, M. Kraft, B.D. Levy, P. Lieberman, S. Luccioli, K.M. McCall, L.C. Schneider, R.A. Simon, F.E. Simons, S.J. Teach, B.P. Yawn, and J.M. Schwaninger. 2010. Guidelines for the diagnosis and management of food allergy in the United States: Report of the NIAID-sponsored expert panel. *Journal of Allergy and Clinical Immunology*. 126: S1–58.

Brown, H. 2009. Validation of cleaning and cross-contact. In *Management of food allergens*, ed. J. Coutts and R. Fiedler, 138–149. Oxford: Wiley-Blackwell.

Brown, H.M., and H.E. Arrowsmith. 2015. Sampling for food allergens. In *Handbook for food allergen detection and control*, ed. S. Flanagan, 181–197. Cambridge: Woodhead Publishing.

Chassaigne, H., J.V. Nørgaard, and A.J. van Hengel. 2007. Proteomics-based approach to detect and identify major allergens in processed peanuts by capillary LC-Q-TOF (MS/MS). *Journal of Agricultural and Food Chemistry*. 55: 4461–4473.

Cho, C., W. Nowatzke, K. Oliver, and E.A.E. Garber. 2015. Multiplex detection of food allergens and gluten. *Analytical and Bioanalytical Chemistry*. 407: 4195–4206.

Cochrane, S., and D. Skrypec. 2014. Food allergen risk management in the factor - From ingredients to products. In *Risk management for food allergy*, ed. C.B. Madsen, R.W.R. Crevel, C. Mills, and S.L. Taylor, 155–166. Oxford: Elsevier.

Cramer, M.H. 2006. *Food plant sanitation: Design, maintenance, and good manufacturing practices*. Boca Raton, FL: CRC Press.

Deibel, K., T. Trautman, T. DeBoom, W. Sveum, G. Dunaif, V. Scott, and D. Bernard. 1997. A comprehensive approach to reducing the risk of allergens in foods. *Journal of Food Protection*. 60: 436–441.

Du, Q., F. Al-Taher, J. E. Schlesser, E. Patazca, C. Pardo, K. Boettcher and L. S. Jackson. 2011. Effectiveness of cleaning regimens for removing milk residue from a pilot-scale HTST processing line. Poster presentation. 2011 Institute of Food Technologists Annual Meeting, New Orleans, LA.

Dunsmore, D.G., A. Womey, W.G. Wittlestone, and H.W. Morgan. 1981. Design and performance of systems for cleaning product-contact surfaces of food equipment: A review. *Journal of Food Protection*. 44: 220–240.

FAARP (Food Allergen Research and Resource Program). 2008. Components of an effective allergen control plan. Available at: http://farrp.unl.edu/allergen-control-food-industry#english Accessed 5 April 2017.

FDA (Food and Drug Administration). 2016. Food allergen labeling and consumer protection act of 2004 (FALCPA). Available at: https://www.fda.gov/Food/GuidanceRegulation/GuidanceDocumentsRegulatoryInformation/Allergens/ucm106187.htm. Accessed 10 May 2016.

———. 2017. FDA Food Safety Modernization Act (FSMA). Available at: https://www.fda.gov/Food/GuidanceRegulation/GuidanceDocumentsRegulatoryInformation/Allergens/ucm106187.htm. Accessed 10 May 2016.

FoodDrink Europe. 2013. Guidance on food allergen management for food manufacturers. Available at: http://www.fooddrinkeurope.eu/publication/FoodDrinkEurope-launches-Guidance-on-Food-Allergen-Management/. Accessed 19 April 2017.

Fu, T.J., N. Maks, and K. Banaszewski. 2010. Effect of heat treatment on the quantitative detection of egg allergens by ELISA test kits. *Journal of Agricultural and Food Chemistry*. 58: 4831–4838.

Fu, T.J., and N. Maks. 2013. Impact of thermal processing on ELISA detection of peanut allergens. *Journal of Agricultural and Food Chemistry*. 61: 5649–5658.

Gabrić, D., K. Galić, and H. Timmerman. 2016. Cleaning of surfaces. In *Handbook of hygiene control in the food industry*, ed. H. Lelieveld, J. Holah, and D. Gabrić, 2nd ed., 447–463. Cambridge: Woodhead Publishing.

Gendel, S.M., and J. Zhu. 2013. Analysis of U.S. Food and Drug Administration food allergen recalls after implementation of the Food Allergen Labeling and Consumer Protection Act. *Journal of Food Protection*. 76: 1933–1938.

Gendel, S., J. Zhu, N. Nolan, and K. Gombas. 2014. Learning from FDA food allergen recalls and reportable foods. *Food Safety Magazine*. April/May edition. Available at: http://www.foodsafetymagazine.com/magazine-archive1/aprilmay-2014/learning-from-fda-food-allergen-recalls-and-reportable-foods/. Accessed 15 September 2016.

Gern, J.E., E. Yang, E.M. Evrard, and H.A. Sampson. 1991. Allergic reactions to milk-contaminated "nondairy" products. *The New England Journal of Medicine*. 324: 976–979.

GMA (Grocery Manufacturers Association). 2009. *Managing allergens in food processing establishments*. Washington, DC: Grocery Manufacturers Association.

Hefle, S.L., and D.M. Lambrecht. 2004. Validated sandwich enzyme-linked immunosorbent assay for casein and its application to retail and milk-allergic complaint foods. *Journal of Food Protection.* 67: 1933–1938.

Holzhauser, T., and M. Röder. 2015. Polymerase chain reaction (PCR) methods for detecting allergens in foods. In *Handbook of food allergen detection and control*, ed. S. Flanagan, 245–264. Cambridge: Woodhead Publishing.

Jackson, L.S., F.M. Al-Taher, M. Moorman, J.W. DeVries, R. Tippett, K.J. Swanson, T.J. Fu, R. Salter, G. Dunaif, S. Albillos, and S.M. Gendel. 2008. Cleaning and other control and validation strategies to prevent allergen cross-contact in food processing operations - A review. *Journal of Food Protection.* 71: 445–458.

Jackson, L. S., and F. Al-Taher. 2009. Comparison of allergen-specific (ELISA) and non-specific (visual inspection, ATP swabs, total protein swabs) methods for detection of soy-based food residues. Poster presentation, International Association of Food Protection (IAFP) Annual Meeting, Grapevine, TX, July 12–15, 2009.

———. 2010. Efficacy of different dry cleaning methods for removing allergenic foods from food-contact surfaces. Poster presentation, International Association of Food Protection (IAFP) Annual Meeting, Anaheim, CA, August 1-4, 2010.

Jones, R.T., D.L. Squillace, and J.W. Junginger. 1992. Anaphylaxis in a milk-allergic child after ingestion of milk-contaminated kosher-pareve-labeled "dairy-free" dessert. *Annals of Allergy.* 68: 223–227.

Laoprasert, N., N.D. Wallen, R.T. Jones, S.L. Hefle, S.L. Taylor, and J.W. Yunginger. 1998. Anaphylaxis in a milk-allergic child following ingesting of lemon sorbet containing trace amounts of milk. *Journal of Food Protection.* 61: 1522–1524.

Lelieveld, H., J. Holah, and D. Gabrić. 2016. *Handbook of hygiene control in the food industry.* Cambridge: Woodhead Publishing.

Levin, M.E., C. Motala, and A.L. Lopata. 2005. Anaphylaxis in a milk-allergic child after ingesting of soy formula cross-contaminated with cow's milk protein. *Pediatrics.* 116: 1223–1225.

Liu, A.H., R. Jaramillo, S.H. Sicherer, R.A. Wood, S.A. Bock, A.W. Burks, M. Massing, R.D. Cohn, and D.C. Zeldin. 2010. National prevalence and risk factors for food allergy and relationship to asthma: Results from the national health and nutrition examination survey 2005-2006. *Journal of Allergy and Clinical Immunology.* 126: 798–806.

Lopez, S. and M. Morales. 2015. Implementing an allergen cleaning validation program: Practical tips. International Food Hygiene Magazine. 26:18-19. Available at: http://www.positiveaction. info/pdfs/articles/fh26_3p18.pdf. Accessed 28 April 2017.

Merin, U., G. Gésan-Guiziou, E. Boyaval, and G. Daufin. 2002. Cleaning-in-place in the dairy industry: Criteria for reuse of caustic (NaOH) solutions. *Lait.* 82: 357–366.

Moerman, F., and K. Mager. 2016. Cleaning and disinfection in dry food processing facilities. In *Handbook of hygiene control in the food industry*, ed. H. Lelieveld, J. Holah, and D. Gabrić, 2nd ed., 521–554. Cambridge: Woodhead Publishing.

Monaci, L., and A. Visconti. 2009. Mass spectrometry-based proteomics methods for analysis of food allergens. *Trends in Analytical Chemistry.* 28: 581–591.

Monaci, L., R. Pilollia, E. De Angelisa, M. Godulab, and A. Visconti. 2014. Multi-allergen detection in food by micro high-performance liquid chromatography coupled to a dual cell linear ion trap mass spectrometry. *Journal of Chromatography A.* 1358: 136–144.

Neogen. 2017. Tree nut allergy test. Available at: http://foodsafety.neogen.com/en/tree-nuts. Accessed 22 May 2017.

Nikoleiski, D. 2015. Hygienic design and cleaning as an allergen control measure. In *Handbook of food allergen detection and control*, ed. Flanagan, 89–102. Cambridge: Woodhead Publishing.

Parker, C.H., S.E. Khuda, M. Pereira, M.M. Ross, T.J. Fu, X. Fan, Y. Wu, K.M. Williams, J. DeVries, B. Pulvermacher, B. Bedford, and L.S. Jackson. 2015. Multi-allergen quantification and the impact of thermal treatment in industry-processed baked foods by ELISA and liquid chromatography-tandem mass spectrometry. *Journal of Agricultural and Food Chemistry.* 63: 10669–10680.

PennState Extension. 2017. Key concepts of cleaning and sanitizing. Available at: http://extension. psu.edu/food/dairy/sanitation-controls/key-concepts. Accessed 27 April 2017.

Powitz, R. W. 2014. Chemical-free cleaning: Revisited. *Food Safety Magazine*. October/November edition. Available at: http://www.foodsafetymagazine.com/magazine-archive1/octobernovember-2014/chemical-free-cleaning-revisited/. Accessed 1 May 2017.

Schmidt, R.H., D.J. Erickson, S. Simms, and P. Wolff. 2010. Characteristics of food-contact surface materials: Stainless steel. *Food Protection Trends*. 32: 574–584.

Schmidt, R. H. 2015. Basic elements of equipment cleaning and sanitizing in food processing and handling operations. University of Florida Cooperative Extension Service, Institute of Food and Agricultural Sciences. Available at: http://edis.ifas.ufl.edu/fs077. Accessed 27 April 2017.

Sheehan, T., J. Baumert, and S. Taylor. 2012. Allergen validation: Analytical methods and scientific support for a visually clean standard. *Food Safety Magazine*. December/January issue. Available at: http://www.foodsafetymagazine.com/magazine-archive1/december-2011january-2012/allergen-validation-analytical-methods-and-scientific-support-for-a-visually-clean-standard/. Accessed 27 April 2017.

Smith, D.L., and J. Holah. 2016. Selection, use and maintenance of manual cleaning equipment. In *Handbook of hygiene control in the food industry*, ed. H. Lelieveld, J. Holah, and D. Gabrić, 2nd ed., 627–648. Cambridge: Woodhead Publishing.

Spanjersberg, M.Q.I., A.C. Knulst, A.G. Kruizinga, G. Van Duijn, and G.F. Houben. 2010. Concentrations of undeclared allergens in food products can reach levels that are relevant for public health. *Food Additives and Contaminants. Part A*. 27: 169–174.

Stone, W.E., M. Jantschke, and K.E. Stevenson. 2009. Key components of a food allergen management program. In *Managing allergens in food processing establishments*, ed. W.E. Stone and K.E. Stevenson, 25–40. Washington, DC: Grocery Manufacturers Association.

Stone, W.E., and J. Yeung. 2010. Principles and practices for allergen management and control in processing. In *Allergen management in the food industry*, ed. J.I. Boye and S.B. Godefroy, 145–166. Hoboken, NJ: Wiley and Sons.

Taylor, S.L., and S.L. Hefle. 2005. Allergen control. *Food Technology*. 59 (40-43): 75.

Taylor, S.L., J.A. Nordlee, L.M. Niemann, and D.M. Lambrecht. 2009. Allergen immunoassays - Considerations for use of naturally incurred standards. *Analytical and Bioanalytical Chemistry*. 395: 83–92.

USDA (United States Department of Agriculture). 2016. Summary of recall cases in 2016. Available at: https://www.fsis.usda.gov/wps/portal/fsis/topics/recalls-and-public-health-alerts/recall-summaries. Accessed 16 April 2017.

Valigra, L. 2010. Integral role for clean-in-place technology. *Food Quality and Safety*. Available at: http://www.foodqualityandsafety.com/article/integral-role-for-clean-in-place-technology/. Accessed 20 May 2017.

van Hengel, A.J. 2007. Food allergen detection methods and the challenge to protect food-allergic consumers. *Analytical and Bioanalytical Chemistry*. 389: 111–118.

Verhoeckx, K.C.M., Y.M. Vissers, J.L. Baumert, R. Faludi, M. Feys, S. Flanagan, C. Herouet-Guicheney, T. Holzhauzer, R. Shimojo, N. van der Bolt, H. Wichers, and I. Kimber. 2015. Food processing and allergenicity. *Food and Chemical Toxicology*. 80: 223–240.

Wilson, D.I. 2005. Challenges in cleaning: Recent developments and future prospects. *Heat Transfer Engineering*. 26: 51–59.

Yeung, J. 2009. Allergen testing and research. In *Managing allergens in food processing establishments*, ed. W.E. Stone and K.E. Stevenson, 57–65. Washington, DC: Grocery Manufacturers Association.

Zhang, X., B. Bedford, T. J. Fu, S. Nekkanti, M. Ross, K. Williams, J. DeVries, B. Pulvermacher, S. Khuda, S. Chirtel, and L. S. Jackson. 2013. Effectiveness of cleaning regimens for removing peanut, milk and egg residue from a pilot-scale cereal bar processing line. Institute of Food Technologists Annual Meeting, Chicago, IL, July 13–16, 2013.

Chapter 9
Survey: Food Allergen Awareness in the Restaurant and Foodservice Industry

David Crownover

9.1 Food Allergies in the United States

According to Food Allergy Research and Education (FARE), researchers estimate that approximately 15 million individuals in the U.S. suffer from food allergies (FARE 2016). According to a report from the CDC, these numbers continue to grow, and the number of children in the U.S. that have been diagnosed with a food allergy increased from 3.4% in 1997–1999 to 5.1% in 2009–2011, a 50% increase (Jackson et al. 2013). These numbers mean that potentially 1 in 13 children has a food allergy.

Individuals that have food allergies must take precautions to avoid foods containing allergens to which they are sensitive. This strict avoidance might be manageable when they are preparing food for themselves, or if a loved one is preparing the food, such as a mother preparing food for a child. However, when these individuals are away from home, this avoidance becomes much more difficult, as they must rely on others to take the same precautions they do in order to ensure the food is safe for them to consume. Dining at restaurants, therefore, poses a significant challenge for those with food allergies, and anyone accompanying them. For instance, if the person with the food allergy has a reaction, their friend, a family member or colleague that dines with them will be impacted by having to respond to the situation.

In an attempt to help consumers make informed choices about food items that they purchase, the Food Allergen Labeling and Consumer Protection Act (FALCPA) was passed into law in 2004. The main goal of this legislation was to address labeling on packaged goods for the 8 major allergens in common terminology (also known as "The Big 8"—peanuts, tree nuts, dairy, eggs, soy, crustacean shellfish, fish, and wheat)

D. Crownover (✉)
ServSafe®, National Restaurant Association, Washington, DC, USA

Microbac Laboratories, Pittsburgh, PA 15229-2132, USA
e-mail: david.crownover@microbac.com

© Springer International Publishing AG 2018
T.-J. Fu et al. (eds.), *Food Allergens*, Food Microbiology and Food Safety,
DOI 10.1007/978-3-319-66586-3_9

to help allergic consumers adhere to an avoidance diet. This labeling requirement excludes foods prepared by the restaurant and foodservice industry. However there are some states that have enacted legislation to require food allergen awareness training of foodservice staff (see Chap. 3) so that they understand how to communicate with an allergic consumer and can accommodate the consumer's needs.

9.2   The Restaurant and Foodservice Industry

The restaurant and foodservice industry in the U.S. is estimated to have had sales of $709 billion in 2015, which is up from $683 billion in 2014, and has experienced an estimated 3–5% growth annually for the last 5 years. It is one of the largest private sector employers, with the industry having almost 14 million employees in 2015, which is up from 12.2 million in 2005. There are over one million restaurants in the U.S. (NRA 2015).

It is estimated that 150 million people go out to eat on a weekly basis in the U.S. According to the National Restaurant Association, the restaurant and foodservice industry garners 47% of the food dollar in the average American household. This statistic is up from only 25% in 1955 (NRA 2015). All of this growth means that more people are dining out than ever before. This growth combined with the continued increase in the number of allergic consumers suggests that more food allergic consumers may develop allergic reactions while dining in restaurants, which is a significant public health concern.

9.3   Restaurant and Foodservice Industry Survey

Because of the increase in allergic consumers and the continued increase in the number of Americans going out to eat, the National Restaurant Association deemed it necessary to review the state of allergen awareness in the food service industry. The results of the review are being used by the National Restaurant Association to develop ways for the food service industry to address this growing trend.

In 2012, the National Restaurant Association contracted a survey that evaluated food allergen awareness and training policies in the restaurant and foodservice industry with Product Evaluations Incorporated (PEI). PEI interviewed managers and owners at 225 restaurants via phone. These restaurants were identified by the National Restaurant News (NRN) as being below the Top 400 restaurant companies in the U.S. and comprised mostly independent owner-operated establishments and small regional chain companies. Those who were interviewed were screened to ensure that they were the purchase decision maker at the unit and were in charge of training staff members on food safety topics. PEI stated that traditionally it is difficult to maintain the interviewee on the phone for more than 15 min, but during this survey, PEI had difficulty keeping the calls shorter than 17 min. This finding

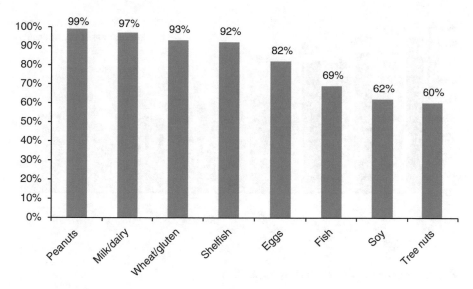

Fig. 9.1 Awareness of specific food allergens, $n = 225$

demonstrated that there might be a passion for the topic and recognition of its importance among the respondents.

In summary, the survey noted that most of respondents from the 225 restaurants surveyed were able to recognize most of the Big 8 allergens and indicated that food allergies were of importance to the restaurant and foodservice industry as a whole. But there was a gap when applying that knowledge to their own operation, both in application and in training staff in food allergy awareness and food allergen control best practices. Here we outline several of the key questions asked in the survey and review the respondents' answers.

Figure 9.1 shows the general awareness and knowledge of the Big 8 allergens in those interviewed through the question, "Which of the following food allergens are you aware of?" There was a strong knowledge of peanuts, milk, and even wheat, but the knowledge decreased, with only 62% and 60% of the respondents being able to identify soy and tree nuts, respectively, as allergens. These figures are similar to those reported by Lee and Xu (2015) in their study on food allergy knowledge among restaurant managerial staff. For instance, they reported that peanuts were correctly identified 93.6% of the time among the survey participants, while soy was only identified correctly 51.8% of the time.

The second question in the survey was, "Which of the following food allergens are important to foodservice in general? Which are important to you/your operation?" Figure 9.2 shows that there were some significant differences in the importance of the allergen to the restaurant and foodservice industry as a whole vs. the relative importance of the allergen to their specific operation. It could be interpreted that many of those indicating that the particular allergens are not important may simply not have recognized that the allergens are being used in their operation.

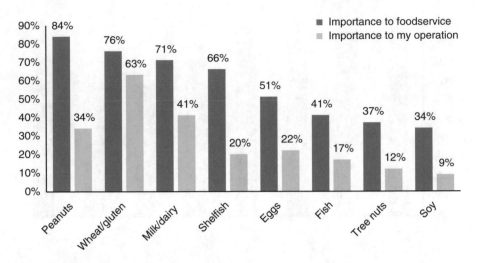

Fig. 9.2 Importance of specified allergens, *n* = 225

However, when looking at the small gap for the wheat/gluten question, we believe that it might actually be a result of the financial importance the individual respondent placed on gluten due to the growing consumer demands for "gluten free" food.

Gluten, while not a true allergen, is a cause of concern for individuals with celiac disease. There is also a consumer movement to avoid gluten consumption as a lifestyle choice. Because of this, many restaurants have begun to provide "gluten free" offerings. The increase in labeling products and menu items as "gluten free" in the market place appears to have led to this smaller gap based on financial and marketing opportunities and not necessarily as recognition of the need to control allergens for food safety reasons. Therefore, we believe that there truly is a gap in awareness and the relative importance that the foodservice operation places on any given allergen.

When asked how important/critical allergens are to the industry, 87% of respondents believed that allergens are very/extremely important to the industry. When asked whether the industry focus on allergens has increased over the past 2 years and will increase over the coming 2 years, 78% and 85%, of the respondents, respectively, answered yes to these questions. It is clear that restaurants expect a continued focus on food allergens and understand that this trend is not going away any time soon. Lee and Xu (2015) confirmed this in their study where 73.6 % of 110 restaurants surveyed agreed that food allergies have become more prevalent, and 71.6% stated that food allergies are a major concern in the restaurant industry.

Yet when asked, "Do you (or your organization) currently train staff on food allergens," only 57% responded "yes" to this question. This result is similar to those observed in the study by Duipuis et al. (2016), which found that of 187 restaurants in Philadelphia, only 52.2% of locations were providing training on food allergen management. Lee and Xu (2015) also demonstrated in their study that those who received food allergen management training had significantly greater knowledge scores (20.2 ± 4.2) about food allergies than those who did not receive training (17.9 ± 5.8, $p < 0.05$). Restaurants having employees with a higher knowledge base can drastically reduce the risk of an error in serving a food allergic customer a product that contains a food allergen.

When probing further with those operations that answered "no" as to whether they train their staff on food allergens, the reasons outlined in Table 9.1 below were given.

While it has been demonstrated that food allergens are important to the industry (87% agree), there appears to be a discrepancy in the recognition for need of training (only 57% train their employees) and awareness on food allergen management in the industry. Even of those survey participants who stated that they are training (57%), the application of the training lacks formality, with only 30% providing "regular" training without defining how frequent that training occurs. The training material is sourced from a wide variety of sources and materials (Table 9.2). It is of concern that for the informal training materials, 40% of the respondents are relying on word-of-mouth stories and 31% on news stories. These two sources may communicate the importance of food allergy awareness but do not provide the structured direction on how to build a food allergen management system.

The final question of the survey was "How do you let your customers know about any allergens in your operations." As shown in Fig. 9.3, an impressive 76% of all individuals surveyed stated that if asked, they tell their customers about the allergens that may be present in the foods they are served. But as identified above, only 57% state that they are training their staff on food allergens. So the question can be asked: Are 19% of restaurants attempting to accommodate individuals with food allergies without properly training their staff on how to do so safely? While there is no requirement to declare food allergens or allergenic ingredients on a restaurant menu, it is concerning that some (1%) of the survey participants think that their facility does not serve any foods with allergens (Fig. 9.3). It is true that every restaurant might not have every allergen, but it is safe to say that every restaurant likely has at least one of the Big 8 allergens in their operations.

Based on this survey, it is clear that there is a gap between overall awareness of food allergens among restaurant operations and the relative importance that is placed on control measures. When this is combined with the low number of companies that responded positively for training (57%) and the potential companies that might be improperly accommodating food allergic consumers without training, it demonstrates that the industry needed assistance and guidance in training to protect public health.

Table 9.1 The reasons for not providing training

Reason	% Answered
I cover the basics/don't think additional training is needed	26%
No reason/never really thought about it	24%
Don't have any training programs/information to use	23%
Don't serve allergen foods/not applicable to us	9%
Don't have the budget for it/costs too much	2%
Don't have time for this additional training	1%

Open ended response to the question "Why don't you conduct training?" ($n = 96$ participants not offering training)

Table 9.2 What information or materials do you use to train staff about food allergens? ($n = 129$)

Formal training materials	Created by me/my organization	54%
	Sourced from outside the organization	30%
Informal training materials	Health Department	49%
	Internet/online resources	49%
	Word of mouth/experience-based stories	40%
	News stories	31%

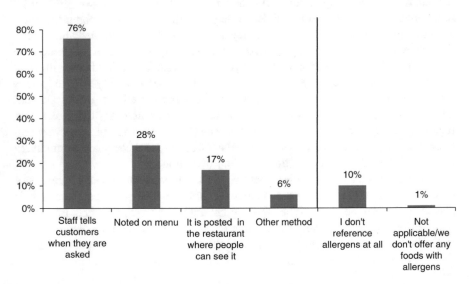

Fig. 9.3 How do you let your customers know about any allergens in your operations? ($n = 225$)

9.4   Development of Training Program for Food Service Staff (ServSafe® Allergens Online Course)

After the survey was conducted, the National Restaurant Association identified that there was a need for a training program for the restaurant and food service industry that focused solely on food allergen awareness and basic food allergen management. The ServSafe® Allergens Online course introduces the student to the basics of food allergies and food allergen management, including the definition of food allergies, recognizing symptoms of food allergies, the dangers of cross-contact, and proper cleaning methods. The course then takes the student through common front-of-house (FOH) and back-of-house (BOH) steps to help minimize and reduce cross-contact. It also instructs the student on communication with the customer and with staff working in other areas of the restaurant, and finally how to deal with an emergency should it occur.

The course has been designed for any employee within a restaurant, including managers, owners, and line staff. It is meant to introduce the students to food allergies and raise their awareness, not as a final step in developing a food allergen management system. As of the writing of this chapter, the course has been identified as an approved training source to meet regulatory requirements in Michigan and Rhode Island.

9.5   Financial Impact of Accommodating the Allergic Consumer

Anecdotally, several restaurant companies have expressed discomfort when addressing whether they should accommodate someone that has a food allergy in their restaurants. Some are worried that if they accommodate for one type of food allergy, then they have to accommodate all of them. Others are worried that it is too difficult to implement a food allergen management system in their operation. Finally, restaurants worry that it is not worth the time and effort to implement a system that will address all of the issues.

In reality, most of these fears are unfounded. Accommodation for one food allergy does not mean that a restaurant has to accommodate all types of food allergies. It is likely not feasible for them to do so based on operational or logistical limitations. Similarly, accommodation does not require a herculean effort or significant capital investment. Many of the steps taken to minimize food safety risks are applicable for minimizing food allergen cross-contact: maintaining good personal hygiene, reducing the potential for cross-contact (similar to cross-contamination), and using effective cleaning and sanitizing procedures.

While there have yet to be any empirical studies that directly address the fiscal impact of implementing a food allergen management program to accommodate allergic consumers, there has been discussion as to the theoretical fiscal impact. For instance, a weekly revenue of $225 million could potentially be generated by U.S. restaurants based on the scenario outlined in Table 9.3.

Table 9.3 Theoretical revenue in the U.S. when accommodating allergic consumers

Est. 15 million individuals with a food allergy
X Theorize 10% can be accommodated (e.g., only a single food allergy like peanuts)
=1.5 million food allergic individuals wanting to dine out
Estimate average family size in the U.S. as ~3 individuals
=4.5 million people (family included) wanting to dine out
Average check is $50
=Potential revenue of $225,000,000 each week

The restaurant and foodservice industry is not alone in the true market potential as it relates to accommodating for the food allergic consumer. The global market for food developed for those with food allergies and intolerances is expected to grow by more than $26.5 billion over the next 5 years. (GIA 2016) These numbers only take into account the potential gross revenue for the food and foodservice industries.

Some have suggested that there is also a potential increase in bottom line dollars when accommodating the food allergic consumer. It is possible for a restaurant to see as much as a 15% increase in profits each year through food allergy accommodation (Jones-Mueler 2011). This is primarily due to the allergic consumer bringing friends or family with them to the restaurant and the distribution of existing overhead to the table. In other words, the more people who are dining at one table, the more profitable that table becomes.

9.6 Challenges to Implementing a Food Allergen Management System and Potential Solutions

A food allergen management system is a set of policies and procedures that a restaurant implements to control and minimize food allergen cross-contact and to clearly and accurately discloses the presence of food allergens. This system is not insurmountable for any restaurant to implement. Yet there are considerations that should be taken into account.

The first and foremost issue, which was outlined in the National Restaurant Association survey, is awareness. While the prevalence of food allergies and general social awareness of food allergies have increased over the past decade, food allergen awareness within the restaurant and foodservice industry has lagged. Restaurants either do not know what they need to do or do not know how to do it. Training on food allergen management can be the first and most important remedy to this challenge.

Unfortunately, the second challenge to implementation directly impacts the solution to the first. Employee turnover in the restaurant and foodservice industry is historically higher than the rest of the private sector. The turnover rate can change depending on the state of the economy. For instance, the entire private sector and restaurant and foodservice industry turnover rates in 2006 were 50.0% and 82.2%, respectively, while those numbers were 42.2% and 62.2%, respectively, in 2013 (NRA 2015). High employee turnover makes continual training difficult. However, the issue may be resolved if key, stable employees are trained and identified as "allergen experts." These "experts" are the go-to people in the restaurant that understand the restaurant's accommodation capabilities and can communicate to the customer as to whether they can be accommodated. They also are essential in ensuring that all the proper allergen management procedures are followed so that safe meals are prepared for the food-allergic customer.

For some full-service and fine-dining restaurants, accurately disclosing allergen information in menus can be a challenge. The menu may not only change weekly, but may change significantly within a day, such as having a salmon special at lunch time and a breaded chicken special for dinner. This also presents an additional challenge through the potential for cross-contact of the salmon to the chicken if they were prepared on the same food preparation surfaces. Controlling them is not difficult with a proper food allergen management system in place. Such a system would require either completely separate preparation surfaces (e.g., a fish-only cutting board), or if that is not possible, a thorough cleaning of the work surface after contact with the salmon and before the chicken comes into contact with the surface.

Finally, space can be a significant challenge to implementing food allergen control measures in restaurants. While food processing facilities might have hundreds and thousands of square feet to use, possibly with completely separate allergen-containing and non-allergen-containing production lines, restaurants are limited by the space they have in the kitchen. Sometimes food preparation space can be as little as six linear feet or possibly less. Yet again a proper food allergen management system can help mitigate this by providing an understanding of how to separate and segregate tasks and cleaning the work space properly.

9.7   Accommodation Considerations

Several states have enacted legislation on allergen awareness training. Massachusetts was the first to enact allergen awareness training with the signing of the Food Allergy Awareness Act in 2009 (The Commonwealth of Massachusetts 2009). The primary goal of this legislation was to require food allergen awareness training of at least one person-in-charge at each restaurant. Rhode Island followed with the Food Allergy Awareness in Food-Service Establishments Law enacted on June 22, 2012 (Rhode Island General Assembly 2012). This new law also requires awareness training in each restaurant. Michigan and Virginia are other states that have also recently enacted similar legislation. These states' legislative initiatives are also discussed in Chap. 3 of this book.

It must be said that accommodation should remain a voluntary option for restaurants to consider based on their business model. A restaurant that specializes in ice cream should not be made to accommodate someone with a milk allergy. It simply may not be feasible for this operation to create a dairy-free product. However, if a restaurant were to voluntarily identify food allergens that they believe they can control, they have then purposefully taken on the task of accommodation and will likely devote more energy and passion to the food allergen management system they should have in place. This will then be evident when the restaurant employee communicates to the customer. Allergic consumers look for a restaurant employee's confident communication as one of the first signs that a restaurant knows or understands how to accommodate a food allergic consumer.

The core of the restaurant and foodservice industry is hospitality. The Oxford Dictionary defines hospitality as "the friendly and generous reception and entertainment of guests, visitors, or strangers." Many restaurants are in operation for the simple reasons that they want to share their talent and give food, comfort, and entertainment to their customers. While eating is a necessary action to sustain life, it has also become a social ritual that is done throughout the day. Individuals with a food allergy are essentially excluded from that ritual because of the very nature of their disease. They must strictly avoid situations where they might potentially ingest a deadly allergen. By identifying the proper steps necessary to safely accommodate food allergic guests, restaurants have an opportunity to reintroduce an underserved and excluded customer base back into this ritual with their friends, families, and colleagues.

References

Duipuis, R., Z. Meisel, D. Grande, E. Strupp, S. Kounaves, A. Graves, R. Frasso, and C.C. Cannuscio. 2016. Food allergy management among restaurant workers in a large U.S. city. *Food Control.* 63: 147–157.

FARE (Food Allergy Research and Education). 2016. Food allergy facts and statistics for the U.S. Available at: http://www.foodallergy.org/file/facts-stats.pdf. Accessed 27 July 2016.

GIA (Global Industry Analysts). 2016. Food allergy and intolerance products - A global strategic business report. Global Industry Analysts. Available at: http://www.strategyr.com/Food_Allergy_and_Intolerance_Products_Market_Report.asp#RCC. Accessed 13 March 2016.

Jackson, K.D., L.D. Howie, and L.J. Akinbami. 2013. Trends in allergic conditions among children: United States, 1997–2011. *National Center for Health Statistics Data Brief* 121.

Jones-Mueler, A. 2011. Interview with Paul Antico, Founder of AllergyEats.com. Available at: http://www.restaurantnutrition.com/Insights/Paul-Antico,-Founder-of-AllergyEats-com.aspx. Accessed 13 March 2016.

NRA (National Restaurant Association). 2015. *Restaurant industry forecast.* Washington, DC: National Restaurant Association.

Rhode Island General Assembly. 2012. Food allergy awareness in food-service establishment. Available at: http://webserver.rilin.state.ri.us/PublicLaws/law12/law12414.htm. Accessed 28 July 2016.

The Commonwealth of Massachusetts. 2009. General laws. Available at: https://malegislature.gov/Laws/GeneralLaws/PartI/TitleXX/Chapter140/Section6B. Accessed 28 July 2016.

Lee, Y.M., and H. Xu. 2015. Food allergy knowledge, attitudes, and preparedness among restaurant managerial staff. *Journal of Foodservice Business Research.* 18: 454–469.

Chapter 10
Allergen Management in Food Service

Miriam Eisenberg and Nicole Delaney

10.1 Introduction

According to the National Restaurant Association 2016 Pocket Fact Book (NRA 2016), nine out of ten consumers enjoy going to restaurants. Half of consumers say restaurants are an essential part of their lifestyle. Seven in ten consumers say their favorite restaurant foods provide flavors that are not easily duplicated at home. Eight in ten consumers say dining out with family and friends is a better use of their leisure time than cooking and cleaning up.

It is likely that a proportion of those restaurant patrons surveyed have allergies. Approximately 4% of adults and 8% of children in the U.S., totaling about 15 million people, have food allergies (FARE 2016a). This allergic population can represent a significant portion of the annual restaurant revenue estimated to be over $700 billion in the U.S. (NRA 2016).

People with allergies or who have family members with food allergies seek restaurants and other food service venues that can safely, cordially, and confidently meet their dietary needs while still serving tasty foods of good quality. These people will bring their families and friends with them to locations that meet these needs and become loyal customers. They will sing the praises for the kitchens that produce a positive dining experience while they may also send out negative comments by word-of-mouth and social media when they feel ill-treated or have had a negative experience. Treating allergic guests well will make efforts worthwhile.

A solid Food Allergen Management Program for a restaurant or for a restaurant chain is good business and should contain:

M. Eisenberg (✉)
EcoSure, a Division of Ecolab, Naperville, IL 60563, USA
e-mail: eisenmir@sbcglobal.net

N. Delaney
Institutional Corporate Accounts, Technical Service, Ecolab, Eagan, MN 55121, USA
e-mail: nicole.delaney@ecolab.com

© Springer International Publishing AG 2018
T.-J. Fu et al. (eds.), *Food Allergens*, Food Microbiology and Food Safety,
DOI 10.1007/978-3-319-66586-3_10

1. Knowledge requirements for the basics on food allergies and allergen control measures
2. Ingredient and supplier management
3. Allergen cleaning
4. Guidelines for service and meal preparation
5. Communication

10.2 Knowledge Requirements for the Basics on Food Allergies

A food allergy is an immune system response in susceptible individuals that typically occurs soon after eating a certain food, generally containing allergenic proteins. Even the smallest amount of the allergy-causing food can trigger symptoms (FARE 2016b), which include:

1. Tingling sensation in the mouth
2. Swelling of the tongue and throat
3. Difficulty breathing
4. Triggering an asthma attack
5. Hives or eczema
6. Abdominal cramps, diarrhea, vomiting
7. Drop in blood pressure
8. Loss of consciousness
9. Death

Some individuals may only experience the mildest symptoms while others may progress through various symptoms of increasing severity. Anaphylaxis is a severe, sudden reaction that can cause a sudden blood pressure drop and a narrowing of airways, thereby preventing normal breathing. No cure exists for allergies. The only way to prevent an allergic reaction is by completely avoiding the allergenic food(s). If an allergen is mistakenly ingested, a medication called epinephrine is often administered in the form of an epinephrine auto-injector carried for emergency use by the allergic person or by medical professionals.

Given the potential severity of allergic responses, a food allergen management program must be well-constructed and clearly administered. The control of allergens and confidence in food preparation are essential to ensuring the safety of consumers with food allergies.

The Food Allergen Labeling and Consumer Protection Act of 2004, referred to as FALCPA (FDA 2004), identified eight food groups which are estimated to cause 90% of allergic reactions in the U.S. Awareness of these allergens has become a top priority. Currently, FALCPA requires *packaged* food items (grocery store items as well as packaged food service products) to be labeled to disclose these ingredients using "plain English." The eight "Major Food Allergen Groups" or "Big 8" as designated by the U.S. Food and Drug Administration (FDA 2004) are:

1) Milk	5) Wheat
2) Eggs	6) Peanuts
3) Fish (i.e., bass, flounder, cod)	7) Tree nuts (i.e., almonds, pecans, walnuts)
4) Crustacean shellfish (i.e., shrimp, crab, lobster)	8) Soybeans

While these "Big 8" are the most common allergens, any food item or ingredient could cause a reaction in susceptible individuals, so even unusual requests may happen. When allergic people dine out, they may request information from that restaurant to assist them in safe choices based on their own needs. Other countries have different lists of common allergens. For example, in addition to the "Big 8" allergens in the U.S., the list of "Priority Allergens" in Canada (Food Allergy Canada 2016) also includes shellfish (i.e., squid, clams, oysters, etc.) as a seafood allergen, sesame seeds, mustard and sulphites.

The European Union Food Standards Agency (FSA 2015) adds celery, sesame seeds, mustard, lupin (a legume related to peanuts), mollusks (i.e., mussels, clams, oysters, scallops), sulfites when >10 ppm in a food or beverage, and cereals containing gluten to their list of common allergens.

Differences in regional common allergens reflect different diet components and the prevalence of different food allergies (milk, eggs, etc.) in different parts of the world. As multiethnic populations merge, common allergens may also change.

Food allergy fatalities must be top of mind in initial planning for food allergen management due to the following facts (Bock et al. 2001):

- 85% of food allergy fatalities occur away from the home with almost half of those deaths occurring due to consumption of a food prepared in a restaurant
- 98% of the fatalities were caused by unintentional ingestion of a food allergen, with peanuts and tree nuts being the predominant causes (39%). A delay in administration of epinephrine was related to the deaths 88% of the time.
- For some allergic people, ingestion of only a minute amount of their allergen can cause an anaphylactic reaction.
- An allergic reaction can happen within seconds or up to an hour after consuming the allergen, and there is high variability in the reactions observed between different individuals.

Considering the health implications to allergic guests when allergens unknowingly enter their diet, it is essential to administer a comprehensive and thoroughly reviewed food allergen management program.

10.3   Establishing a Food Allergen Management (FAM) Program

In establishing a FAM Program, consider it an educational as well as management project. Federal and local regulations should be taken into consideration both for guidance and to be in compliance. Allergen knowledge for recipes is dependent on proper labeling as well as communication with vendors. Well-written recipes along with training for food handlers can prevent cross-contact of allergens. Cleaning is also integral to allergen management. All aspects are pulled together through communication and training for each associate's role in keeping the allergic guest safe.

10.3.1 Regulatory Considerations

Once the determination has been made to proceed with development of an FAM program, regulatory requirements should be reviewed.

As of this writing, specific labeling is not required on restaurant menus highlighting the presence of allergens, nor are allergen-managed[1] options required in the U.S. This is unlike the EU where since December 2014, the EU Food Information for Consumers Regulation No. 1169/2011 requires food businesses to provide allergy information on food sold unpackaged, in for example catering outlets, deli counters, bakeries and sandwich bars (FSA 2015). While some chains and individual restaurants may offer some allergen-managed items, such offerings are voluntary. Restaurant owners or national chains must take into consideration that for some allergic individuals even a trace amount of an allergen can cause a severe life-threatening reaction (anaphylaxis) and be prepared to further train on emergency measures for potential allergic reactions.

Currently the direction given in the U.S. FDA Food Code (FDA 2013) for food service operations is focused on manager knowledge as well as employee training. In the 2005 U.S. FDA Food Code, Part 2-1 addressing "Supervision," an addition was made to the subpart 2-102 "Demonstration of Knowledge." This new subparagraph §2-102.11(C)(9) states "The person in charge shall demonstrate knowledge by…describing FOODS identified as MAJOR FOOD ALLERGENS and the symptoms that a MAJOR FOOD ALLERGEN could cause in a sensitive individual who has an allergic reaction." This means that there is an expectation for managers to know the eight major allergens and symptoms of an allergic reaction in order to be able to respond appropriately such as by calling for emergency assistance. It is important that "9–1–1" is called when a guest is in distress since symptoms can be progressive.

Added in 2009 to the U.S. FDA Food Code in subpart 2-103 "Duties of the Person in Charge," paragraph §2-103.11(L) states that "the person in charge shall ensure that … EMPLOYEES are properly trained in FOOD safety, including food allergy awareness, as it relates to their assigned duties." While the interpretation of this may vary with the local health department and/or restaurant, it can range from how servers answer customers' questions, to communication to the preparation staff to ensure that the dishwasher properly cleans all the wares.

---

[1] The term "allergen-managed" is used as opposed to "allergen-free" since under most circumstances, food service kitchens are not free of allergens but special management is put into place to minimize allergen cross-contact.

Since the U.S. FDA Food Code is a set of guidelines to food regulatory jurisdictions rather than laws, the adoption and enactment of any component of the U.S. FDA Food Code are highly variable across the country. Many states have adopted Food Code recommendations requiring manager knowledge of allergens and some employee-related training (see Chap. 3 of this book). The Commonwealth of Massachusetts has set higher standards. It is required that managers with food protection certification also complete allergen-awareness training through a program recognized by the Massachusetts Department of Public Health. The Massachusetts Health Department approved a Massachusetts Allergen Training program, which includes a 30-min online training video developed by Food Allergy Research and Education, Inc. (FARE). In addition, a notice is required on menus or menu boards that it is the customer's obligation when ordering to inform their server about any food allergies.

10.3.2   Accurate Ingredient Information

To set up a Food Allergen Management (FAM) Program for a restaurant or chain, current and correct ingredient information must first be established. Standard recipes must be reviewed, and those without certain allergens can be used as already created. Next is the choice of either modifying recipes to remove and/or replace common allergens with other ingredients or to only use current menu items that are already made without certain allergens.

It is important to recognize that this will be an allergen "management" or "control" program and not an "allergen-free" program unless the choice is made to eliminate certain allergens from the kitchen. The reality is that most food service kitchens use numerous ingredients; hence, allergens are likely present in the environment.

Complete knowledge of all ingredients is imperative so an inventory must be created. All ingredients must be in compliance with FALCPA by identifying the eight major food allergens on food labels in understandable language. For example, the specific type of nut (almond, pecan, walnut, etc.) must be listed, or the exact type of crustacean or fish must be indicated. Soybean, soy, and soya may be used interchangeably. When any of the major allergens are present in flavorings, coloring and additives, they must also be clearly indicated. While FALCPA does not apply to alcohol, fresh meats, poultry, and certain egg products, the USDA Food Safety and Inspection Service (USDA-FSIS 2013) encourages the use of allergen statements that align with FALCPA.

10.3.3   Working with Product Labels and Ingredient Suppliers

When developing a FAM program, it is essential to communicate expectations to ingredient manufacturers and supplies. Working with food suppliers to receive accurate allergen information is vital. Assumptions cannot be made about ingredient contents, since some allergens may not be obvious. For example, most soy sauce and commercial broth concentrates contain wheat. Frozen French fries are often dusted with dairy or wheat components. Some suppliers will have similar products with differing ingredients. It is important to check labels since suppliers are required to follow FALCPA.

Along with allergen labeling, there may be advisory statements on the labels regarding the manufacture of food ingredients on shared equipment or in a facility that processes allergen-containing products. Examples of advisory statements include wording on the label that a product "may contain," is "processed in a facility that …," or is "manufactured on shared equipment with …" an allergen. These statements are not consistent nor are they regulated. Items with these labels may pose an allergen risk and require management, so substitutions with allergen-free items for these ingredients are recommended. For example, cookies "processed in a facility that also processes products with peanuts" should be replaced with a cookie without such a warning.

In addition, it is essential for suppliers to provide notification when reformulating ingredients, particularly when these reformulations change the allergen content. Distributors should also be included in communication regarding ingredient consistency and should be required to notify a facility of any ingredient substitutions. Even with these arrangements, labels should be routinely checked on the ingredients used for allergen-controlled items. Cooking staff should be trained to double-check ingredients that they may only occasionally utilize for special menu items.

10.3.4   Managing Your Ingredients

To assist in managing recipe ingredients for allergens, a spreadsheet can be utilized which lists every ingredient used in each recipe. This spreadsheet can then be cross-referenced with the "Big 8" allergens indicating the allergens present in each ingredient, and therefore, in specific menu items. With the information from ingredient labels in conjunction with recipes, a decision can be made to modify recipes for allergen-free menu items by either substituting other ingredients or by sourcing allergen-free ingredients (see Table 10.1 for an example). Using a tool such as this, decisions can be made regarding what menu items are already suitable for an allergen-free offering and also what modifications could be made to eliminate certain allergens to create new offerings.

10.3.5   Facility and Equipment (FAM)

As part of an FAM, dry storage areas, coolers and serving lines need to be organized. Food and ingredients that might be used for allergen-managed menu items

Table 10.1 Allergen spreadsheet example

Ingredients	Dairy	Soy	Peanuts	Tree Nuts	Eggs	Wheat	Fish	Crustaceans
Bob's soy sauce		X				X		
Bob's gluten-free soy sauce		X						
Worcestershire sauce		X				X	X	
Bleu cheese	X							
Unbleached flour						X		
Corn tortillas								
Wheat tortillas						X		
Pre-breaded shrimp		X				X		X

should be clearly labeled with allergen content once removed from original packaging and also appropriately stored to prevent contamination of allergen-free foods with allergen-containing foods. For examples, chopped nuts should be segregated from other ingredients in order to limit contamination of other foods.

You must have adequate equipment and other wares to offer allergen-managed items. In some cases, you may need duplicate equipment to successfully manage your FAM Program. Readily available clean, extra wares and equipment allow for quick preparation and the need to avoid cross-contact.

10.4   Allergen Cleaning

Overall cleaning and cleanliness aspects are integral to the FAM plan's success and involve many staff members. An allergen, often a protein, differs from foodborne pathogens, which are tiny living organisms such as viruses or bacteria. The proteins that elicit an allergic response are not living; therefore, use of sanitizers alone is not an effective allergen management tool. Allergen management must involve physical removal of the allergenic protein through thorough cleaning of a surface or area.

10.4.1   Five Factors for Successful Cleaning

Proper cleaning comprises the successful integration of the following factors: time, temperature, chemical action, mechanical action, and procedures. The inclusion of each of these components is necessary for the removal of all soils from a surface, including allergens. Different surfaces within the foodservice establishment will require separate processes, temperatures, and chemical choices. It is important to specify all requirements in a procedures document and train responsible associates on proper execution.

Time needed to properly execute cleaning of a surface will vary. This factor also includes the frequency with which the cleaning should be performed. Temperature requirements vary for different types of food soils. Higher temperatures may be achieved by using an enclosed mechanical warewashing machine which limits the chemical and temperature exposure of employees. Chemical action specifies the proper chemical choice and concentration for cleaning the specified surface and type of food soil. The products chosen, including cleaning tools, should be compatible with the surface. Incompatibility may lead to surface deterioration which may become a harborage for soil and hence allergens. Mechanical action refers to pressure applied to a surface, either through mechanized water pressure or manual pressure through cleaning implements. Implements can include brushes of various sizes, scrubbing pads, squeegees, or other specialized devices designed for cleaning. Procedures must define the steps taken to achieve a clean and sanitary surface. Ensuring proper execution of the procedures can be done through training and observation, as well as posting of placards and visual aids. Another aspect to consider is that the necessary tools should be cleaned and maintained so that they do not contribute to cross-contact and stored only with tools used to clean food contact surfaces.

10.4.2   Detergent Considerations

It is important to choose the correct detergent ingredients. There are four main types of soil: fats, carbohydrates, proteins, and minerals. Each has a different composition, and removal can be aided by different chemical components. Some specialty detergents will have a combination of ingredients targeted at removing multiple classifications of soils. Most foods will contain more than one soil type. Take some time to identify what types of food soils are present in different areas of your establishment.

There are six main chemical components of a cleaning detergent or cleaning chemical that aid in removing soil alkaline, acid, oxidizer, enzyme, solvent, and surfactant (Table 10.2).

Alkalines are characterized as chemicals having a pH > 7. Inclusion of alkaline materials in a formula aids in dissolving or dispersing any organic soils. Cleaning chemicals containing alkaline materials are used if soils contain carbohydrates, proteins, or fats. Acids are identified as chemicals having a pH < 7. Acids aid in dissolving mineral soils, such as those responsible for water hardness. If the water source used in a restaurant has significant hardness, consideration should be given to periodically using acid cleaner to minimize scale buildup on surfaces, especially those exposed to water for long periods of time. Examples include steamers and the warewashing machine area.

Oxidizers liberate oxygen to hydrolyze larger molecules. Two common types of oxidizers are sodium hypochlorite (bleach) and peroxide. Use of oxidizers is suggested with protein soils. The oxidizers in the peroxide or bleach chemicals attack

Table 10.2 Chemical components of detergent for different types of soil

Chemical component	Mechanism	Example	Soil type
Alkaline	Dissolve or Disperse	NaOH, KOH	Fats, Carbohydrates, Proteins
Acid	Dissolve	Phosphoric, Nitric	Minerals
Oxidizer	Hydrolyze	Bleach, peroxide	Proteins
Enzyme	Hydrolyze	Protease	Proteins
Solvent	Dissolve	Water, alcohol	Fats, Carbohydrates
Surfactant	Liquefy, Emulsify	Fatty acids, Detergents	Fats, Carbohydrates

oxygen sources within the protein to chemically hydrolyze the protein into smaller pieces, aiding in soil removal. It is easier to dissolve smaller protein fragments (peptide) segments in water, and easier for the surfactants to keep the peptide pieces in the water solution to be carried away instead of redepositing on the food contact surfaces.

Another way to hydrolyze proteins is through use of a proteolytic enzyme. Proteolytic enzymes are used in many cleaners where chemicals having high pH or harsh oxidizers are not recommended. Enzymes can be used at room temperature and are not damaging to many surfaces. Solvents are used to dissolve another substance. Water is the most common solvent, although alcohols are occasionally used. Inclusion of an alcohol-based solvent can be useful in removing fats, which do not dissolve well in water. Some carbohydrate soils can also benefit from use of solvent additives. Surfactants are compounds that reduce the surface tension between components in a liquid solution. They are commonly used to emulsify fats, oils, and carbohydrates in water solutions, which make them less likely to redeposit on the cleaned surface. The addition of surfactants in cleaners allows for use of lower temperatures when cleaning surfaces, which is especially useful when cleaning large amounts of fats and oils. Fats and oils, on the other hand, may need to be liquefied to hasten removal. This is usually done by raising the temperature of the wash water, or by cleaning a surface (e.g., grill) while it is still warm. Increasing the temperature allows the fats to melt from a solid to liquid form, which aids in cleaning from the surface.

10.4.3   Methods of Cleaning

Many foodservice operations have a variety of methods available to clean equipment exposed to allergens. Some, like wares, can be processed through a warewashing machine. It is especially important to follow proper procedures when preparing the wares for the mechanical washing process. This includes precleaning to remove any visible soils and rinsing before properly loading the dish rack. This minimizes the amount of soil that is introduced into the wash tank of the warewashing machine.

High levels of soil in the wash cycle can lead to redeposition of soil, including allergens, on subsequent racks of dishes. For recirculating machines, it is recommended to change the water at least every 2 h. It may also be recommended to wash allergen-sensitive items (i.e., utensils used primarily for allergen-managed procedures) right after changing the wash water to minimize the potential for soil redeposition. It is also important to ensure that employees are trained regarding which items should not be placed through the warewashing machine. This may be for material compatibility, temperature constraints, or size constraints. Large pots, for example, may shield the wash arms from reaching all surfaces and lead to allergens remaining on the surfaces. Allergens not removed during cleaning and sanitizing can be a potential source for cross-contact in storage, preparation, or use.

Other operations utilize manual methods of cleaning wares. As with mechanical ware washing, it is important to follow proper prewash procedures to minimize the level of soil in the wash basins. It is also a good practice to follow established cleaning procedures for any implements such as brushes used to remove soil from the ware. Some cleaning implements have the ability to harbor food soil which can cause both microorganism and allergen risks. In manual environments, it is also a good practice to wash allergen-sensitive items first or directly after changing the basin solutions. Again, this will help minimize the potential for cross-contact of allergens onto wares. As previously mentioned, sanitizing wares alone is not adequate to remove allergens without thoroughly cleaning all surfaces first. Employees should be reminded to visually inspect items emerging from both the warewashing machine and manual washing for presence of food debris and rewash these items.

10.4.4   Cleaning Beyond the Back of House

Environmental cleaning can also help reduce the risk of allergen cross-contact. Reducing environmental soil loads on walls, equipment, and any non-food contact surfaces incorporates many of the recommendations already mentioned: selecting proper tools, utilizing the proper procedures, and training staff on which combinations are most effective on different surfaces. Environmental cleaning is usually performed manually, though some equipment may be disassembled for mechanical cleaning. When analyzing allergen risk, focus on environmental surfaces in close proximity to where allergen-free ingredients or meals are prepared or stored. Pay special attention to the undersides of exposed equipment, hinges or sandwich points, or other areas that are often overlooked. Listing these surfaces on current cleaning checklists and training staff on procedures are ways to ensure these surfaces are free of soil, allergens and other contaminants. Another practice is to place these particular areas on a rotating sanitation schedule to focus on a specific area every shift or other unit of time. This allows more detail to be placed on areas of concern and allows all environmental surfaces to be cleaned on a rotating basis.

Allergen cleaning should also be considered at the front of house, including managing "touch points," areas or items that the allergic guest may touch. Since wiping down surfaces like tables, bar tops and chair backs with a sanitizer alone, a common practice, does not remove allergens, a proper procedure of washing with a soapy solution and rinsing with clean water with a final sanitizing step is possible when the allergic guest makes an advanced reservation at the restaurant. However, if a food service facility were to utilize this practice as part of normal procedure, they could always be ready for the allergic guest. Changing table linens between guests also can be used to reduce exposure of guests to allergens from table surfaces.

10.5 Service and Meal Preparation

Once determinations are made about what menu items are available or will be modified for your allergen-controlled menus, guidelines for service and preparation must be created.

As part of this, it is important that facts are laid out for all associates in case there are any misbeliefs. In a 2007 study (Ahuja and Sicherer 2007), restaurant and food establishment personnel indicated that 58% of their establishments offered no form of food allergy education. Other findings of this study included the following:

- 70% of those surveyed **felt that they could guarantee a safe meal**
- 24% thought that consuming **small amounts** of allergen is safe
- 35% believed that fryer **heat can destroy** allergens
- 54% thought that a **buffet is safe** if kept **clean**
- 25% of those surveyed thought that it is **safe to remove an allergen from a finished meal**

Overall, the study indicated that many beliefs were erroneous and that the comfort level was greater than the knowledge level.

10.5.1 Service and Front of House Operations

In terms of service, it is essential to decide how front-of-house staff will be involved. Starting with the host stand and the reservation system, guests can be asked in advance if there are any special dietary needs. This will allow a restaurant with a FAM to be alerted that a guest with special dietary needs will be coming in at a specific time. Fresh linens for a table will also eliminate allergens on the table surface.

While some sources of allergens may seem obvious for some menu items, such as cheese which contains milk or croutons which contain wheat, if a server is unaware of "hidden" ingredients, the FAM is not being properly implemented. Program knowledge and proper training are important. When managers, chefs or trained servers take the order for the allergic guests, it is important that offerings and their preparation are clearly communicated. Special notations should be made on the order slip and entered into a computerized order system, if available. Ordering systems can be designed to highlight allergies and special preparation requirements so that correct written confirmation of the special order arrives at the cook's line. It is also ideal to verbally communicate these special orders directly to the kitchen staff, whether to a lead chef/cook or to the line manager. Each staff member who handles the order must be aware of the allergy and communicate with each other—from reception to the server to the kitchen and then back to the server.

If a restaurant can offer thoughtful amenities to the allergic guest, this helps build program satisfaction for the guests as well as develop a relationship between the guest and restaurant brand. For instance, if a guest identifies a dairy allergy and normally butter is placed on the table, that guest might appreciate being offered olive oil as an alternative. If a guest indicates that they have a wheat allergy, the staff may ask whether they want bread placed on the table or perhaps crudité can be substituted.

10.5.2   Back of House Operations

Once the order is communicated to any line cook or chef, there must be a break in the preparation action. This is to allow for communication and to prevent cross-contact—the carryover of an allergen from one surface to another.

Whether space allows a separated designated preparation area to be kept as an allergen-controlled zone or an area must be cleaned for preparation, the responsible person must get the area and any needed equipment ready. Food contact surfaces must be first washed and rinsed before they are sanitized since sanitizing alone is not sufficient to remove potential allergen present. Hands must be washed and fresh gloves worn for ready-to-eat food handling. Clean utensils, dishes, bowls, etc. must be used for the allergen-controlled food items. In some operations, a special "allergen-control" kit is kept clean for ready use. Commercial allergen-control kits are often purple, an unusual but visible kitchen color, and may include a cutting board, knife, tongs, scoop, spatula and thermometer. Included wares can be customized for the operation. A kit is not required but can be a convenience. Instead of a commercial allergen-control kit, needed wares for a given operation can simply be put aside in a covered container for convenient use. Care must be taken in cleaning and sanitizing items in the kit before and after each use, as meals that require the use of dedicated equipment may have different requirements (i.e., dairy- and tree nut-free versus dairy-free).

Supervision of the staff preparing the allergen-controlled items can help assure accuracy and proper technique. Routine training for all staff about sources of cross-contact and prevention is essential. If any errors are made during preparation, the menu item must discarded and be completely remade. Picking off the allergen, such as croutons on a salad for a wheat allergic guest, can leave traces of the allergen which may be lethal for the highly allergic guest. Thus, removing allergens is not an acceptable practice.

For cooking surfaces like open-fire grills and griddle surfaces, it should be ascertained if any common allergens are ever cooked on those surfaces. In some operations, meats, poultry, fish and crustaceans are cooked on different areas of the grills. If not, some areas are kept uncontaminated from certain allergens. In some cases, items will be cooked over the grills in clean pans or over clean foil. Each facility must be assessed individually.

Additionally, care must be taken in open kitchens. Allergic reactions do not only occur from ingestion, but can also occur from inhalation. Reactions from inhalation of fish, egg, legumes, buckwheat, and milk associated with active cooking have been documented (Roberts et al. 2002). For operations where dusting surfaces with flour or cooking on open grills are common, allergic reactions can occur, so allergic guests should be seated away from these types of processes.

For each step of creating a plate for the allergic guest, washed hands should be covered with new disposable gloves. Once the allergen-managed dish has been plated, the line manager should double-check the plate, and have it carried separately to the guest. If it must be carried on a tray with other food or by hand, care should be taken to not contaminate the allergen-managed dish with contents of other plates. Some operations use color flags in the special food items to help identify them along with the server acknowledging the specially prepared item when placed in the front of the guest. Reassurance at each stage is a comfort to the guest.

10.6   Communication

Good communication cannot be over-emphasized. Information-sharing at every team member position helps to minimize risk of allergen cross-contact. Communication skills must be coupled with management of staff behaviors such as properly answering allergen questions and preparing special meals. There is no room for error when a life could be at risk. When the FAM includes everyone involved in the food service establishment, the interactions with the guest, the flow of the food order, the preparation and service steps along with proper cleaning behaviors, confidence is instilled with both employees and guests.

10.7   Summary

Setting up an all-encompassing Food Allergen Management (FAM) Program is a large endeavor which requires input and cooperation from many sources. General knowledge on basics of food allergies, common food allergens, reaction symptoms, inherent risks and labeling guidelines are starting points. Complete understanding of the life-threatening risks of a poorly executed program must be acknowledged as part of the commitment offering an allergen-managed special menu.

Planning must take regulatory requirements into consideration along with integration of thorough ingredient information. Facility aspects, proper food storage and equipment availability are aspects of consideration for the flow of food that when well-managed will limit the potential of cross-contact and problems for the guest. Proper implementation of a cleaning and chemical program is a vital adjunct to the FAM program. Of course, the human aspects of training and communication must be continually emphasized with regular program, menu and implementation review. Proper management along with good corporate communication and support for chain food service facilities will help in offering allergic guests a safe and secure dining experience, and this level of comfort can help establish long-term relationships with these consumers.

References

Ahuja, R., and S.H. Sicherer. 2007. Food allergy management from the perspective of restaurant and food establishment personnel. *Annals of Allergy, Asthma & Immunology*. 98: 344–348.

Bock, S.A., A. Munoz-Furlong, and H.A. Sampson. 2001. Fatalities due to anaphylactic reactions to foods. *Journal of Allergy and Clinical Immunology*. 107: 191–193.

Food Allergy Canada. 2016. Food allergens. Available at: http://foodallergycanada.ca/about-allergies/food-allergens. Accessed 28 September 2016.

FARE (Food Allergy Research and Education). 2016a. Facts and statistics. Available at: http://www.foodallergy.org/facts-and-stats. Accessed 1 September 2016.

———. 2016b. Symptoms. Available at: http://www.foodallergy.org/symptoms. Accessed 1 September 2016.

FDA (Food and Drug Administration). 2004. Food Allergen Labeling and Consumer Protection Act of 2004 (Public Law 108-282, Title II). Available at: http://www.fda.gov/Food/GuidanceRegulation/GuidanceDocumentsRegulatoryInformation/Allergens/ucm106187.htm. Accessed 28 September 2016

———. 2013. Food Code 2013. Available at: http://www.fda.gov/food/guidanceregulation/retail-foodprotection/foodcode/ucm374275.htm. Accessed 26 September 2016.

FSA (Food Standards Agency). 2015. Food allergen labelling and information requirements under the EU Food Information for Consumers Regulation No. 1169/2011: Technical guidance. Available at: http://www.food.gov.uk/sites/default/files/food-allergen-labelling-technical-guidance.pdf. Accessed 28 September 2016.

NRA (National Restaurant Association). 2016. 2016 Restaurant industry pocket factbook. Available at: https://www.restaurant.org/Downloads/PDFs/News-Research/PocketFactbook2016_LetterSize-FINAL.pdf. Accessed 1 September 2016.

Roberts, G., N. Golder, and G. Lack. 2002. Bronchial challenges with aerosolized food in asthmatic, food-allergic children. *Allergy*. 57: 713–717.

USDA-FSIS (United States Department of Agriculture - Food Safety and Inspection Service). 2013. Journal of allergy and clinical immunology. Available at: http://www.fsis.usda.gov/wps/portal/fsis/topics/food-safety-education/get-answers/food-safety-fact-sheets/food-labeling/allergies-and-food-safety/allergies-and-food-safety/. Accessed 1 September 2016.

Chapter 11
Managing Food Allergens in Retail Quick-Service Restaurants

Hal King and Wendy Bedale

11.1 Introduction

For the first time, Americans now spend as much money on food away from home as they do at home (USDA-ERS 2015), with about 30% of food money spent away from home going to limited-service restaurants including fast-food (also known as quick-service) restaurants (USDA-ERS 2016). The growing number of consumers dining out contributes to the fact that restaurants are the location where a substantial percentage of anaphylactic reactions occur due to food allergens. A 2002 study with mostly North American participants found that 17.6% of food-induced allergic reactions occurred in restaurants (Eigenmann and Zamora 2002). In a more recent study, 12% of food-induced anaphylaxis cases resulting in pediatric emergency room visits were found to occur following a restaurant visit (Rudders et al. 2010), while in adults, more than half of food allergen exposures leading to emergency room visits occurred at restaurants (Banerji et al. 2010). Food allergic reactions that occur in restaurants can have serious outcomes: an analysis in the U.S. found that nearly half of food allergy deaths recorded in a 1994–2006 registry were associated with food establishments, with some occurring at quick-service restaurants (Weiss and Munoz-Furlong 2008).

Management of food allergens in quick-service restaurants has significant differences compared to food allergen control in full-service restaurants. In some cases, independent and franchised quick-service restaurants have additional challenges to control food allergens and protect food allergic individuals. Key differences between

H. King (✉)
Public Health Innovations, LLC, Fayetteville, GA 30215, USA
e-mail: halking@pubhealthinnovations.com

W. Bedale
Food Research Institute, University of Wisconsin-Madison, Madison, WI 53706, USA

© Springer International Publishing AG 2018
T.-J. Fu et al. (eds.), *Food Allergens*, Food Microbiology and Food Safety,
DOI 10.1007/978-3-319-66586-3_11

quick-service restaurants and full-service restaurants with respect to food allergen risk and controls for managing such risks are introduced in Table 11.1.

Other chapters in this volume (Chaps. 9 and 10) discuss food allergen management and control measures more relevant to full-service restaurants. This chapter will illustrate some of the risks related to food allergens that are associated with quick-service establishments, while also highlighting food allergen management methods (including supplier control, restaurant operation management, and ingredient communication) appropriate for quick-service restaurants.

11.2   Food Allergen Risks at Quick-Service Restaurants

11.2.1   Fast Food Is Packaged for Immediate Consumption

Food prepared and served in restaurants, especially quick-service restaurants, is normally produced and packaged for immediate consumption. Because of this service model, most packaging of quick-service restaurant foods does not list product ingredients, nor include any precautionary or advisory statements that the food may contain undeclared allergens. This is different from packaged food from a retail store where customers can review the declared ingredients in the product and precautionary statements about how the product was produced (e.g., manufactured in a facility that also processes tree nuts) and make a decision before they consume or serve the food.

Likewise, while manufactured and packaged foods can be tested for the presence of undeclared allergens prior to distribution, there is neither time nor appropriate technology in a restaurant to conduct any kind of food allergen testing after the final food product is prepared and before a customer eats it (other than a cursory visual inspection (e.g., whole nuts) which would not detect unseen allergens like milk or egg protein in a coating).

The emphasis on rapid production, maintaining food at hot or cold temperatures, and delivery of food for quick-service can result in order mix-ups, including (and particularly) when special orders are made to a regular menu item in response to an allergic customer's request. For example, a special order sandwich might be placed in a heated tray with other sandwiches to keep it warm while the rest of the order is made, which might result in delivery of the wrong sandwich to the customer. While some special orders may simply be a matter of preference (e.g., no pickles) and would not require the attention that food allergen-related special orders might need, quick-service restaurants should consider handling all special orders in the same manner to prevent potentially serious mistakes. Likewise, because the packaging for handling and holding foods for customers is oftentimes preprinted, some restaurants use labels (stickers) placed on the food packaging to indicate a menu item has been specially prepared (e.g., without egg). A caution in this however is that such a label might lead a customer to mistakenly believe that the product was prepared to prevent all egg cross-contact although this may not be the case.

Table 11.1 Differences in food allergen risks and controls between quick-service and full-service restaurants

Category	Quick-service restaurant	Full-service restaurant	Food allergen risks and controls
Managing supply chain and ingredients	Limited menu, standardized across many restaurants with less frequent change	More extensive menu with more frequent changes	A more limited menu with fewer ingredient changes can make food allergen control easier.
	Menu items have very defined ingredients	Chefs may not want to divulge recipe ingredients and may not make the same dish the same way each time	More control and transparency over recipe ingredients may be possible in a quick-service restaurant chain.
	Ingredients for multiple restaurants usually supplied from a central location	Ingredients more likely to be locally sourced, and source may change often	Supply chain can be more tightly controlled in a quick-service restaurant; however, if a problem occurs, more people may be affected.
Restaurant operations	Menu items may be fully or partially prepared at a central location rather than at the restaurant	Menu items prepared in an often crowded kitchen within the restaurant	A central preparation facility or the use of prepared foods from a manufacturer may result in more space and management systems to better segregate preparation of allergen-free recipes.
	Very frequent staff turnover	Less frequent staff turnover	Training on food allergens present in menu items and communications with customers may be harder to conduct and keep up to date when staff turnover is high.
	Usually part of multi-restaurant franchising chain	May or may not be part of a restaurant franchising chain	Restaurants that are part of chains do not have to create their own food allergen management plan from scratch and may benefit from regular inspections and food safety management systems provided by the parent company. The parent company of a large chain may have more ability to audit and inspect ingredient suppliers than an independent full-service restaurant.
Communication with customer	Food may be served in a wrapper or other single-use package	Food is served on a plate	Wrappers or packaging can be labeled or tagged to warn of food allergen presence.
	Order process is very fast	Ordering process is slower	Less time is available at a quick-service restaurant for customers to convey food allergy concerns or ask questions and receive information when ordering.
	Technology is generally used in ordering and communicating order to kitchen	Communication of order to kitchen is often done manually	The opportunity to automatically report ingredient and food allergy information to the customer and the kitchen may be greater at a quick-service restaurant.

11.2.2 Undeclared (and Unexpected) Food Allergens May Be Present

Undeclared and unexpected food allergens are hazards to anyone with food allergies when dining out. Menu changes, changes in ingredient sources, manufacturing defects, and cross-contact during food preparation can all introduce new food allergen hazards at a quick-service restaurant.

In a quick-service restaurant, menu changes may not occur with as much frequency as in other types of restaurants, but it is important that even "limited-time" products be made with the same attention to food allergens as menu staples. "Limited-time" products might utilize ingredients from new suppliers who have not yet been fully audited, or menu allergen information may not be updated promptly when changes or additions are made. Similarly, changes in ingredient suppliers for regular menu items have the potential to introduce unexpected allergens. If the source of an ingredient is changed and the new source has not been thoroughly investigated or audited by the restaurant, food allergens may be unknowingly introduced into the restaurant's products. Manufacturing defects, errors in disclosure of allergen information, or errors/gaps in label control can occur at the supplier facility. An ingredient could be accidentally contaminated with traces of peanut if it is produced in a facility that also handles peanuts, for example. Such a problem at a quick-service restaurant chain could impact many restaurants that are part of the chain, ultimately affecting a large number of customers.

Cross-contact with allergens can occur during food preparation at the restaurant, especially in the small and crowded spaces associated with quick-service restaurant kitchens. In some cases, no segregated space for allergen-free food preparation may be available in the kitchen. Risk of allergen cross-contact can also occur with special orders items. For example, an egg-allergic customer may ask whether a breakfast sandwich is available egg-free. The employee at the counter might say "yes" and simply remove the egg from an already assembled sandwich or make a sandwich without the egg using the same gloves that were just used to make egg-containing sandwiches.

11.2.3 Customers May Be Unable to Obtain Food Allergen Information

Food allergen risks may also arise from gaps in communicating allergen information to consumers due to consumer behavior, lack of employee knowledge, failure to establish food allergen inventory, or not making allergen information available to employees or consumers (e.g., failure to disclose allergen in menu). Consumers with food allergies may be hesitant to ask for information about food allergens at restaurants due to social considerations such as embarrassment or fear of being perceived as picky (Leftwich et al. 2011). In a quick-service restaurant, the often rushed (especially during drive-thru ordering) and rather public ordering process has the potential to increase this reticence.

Even if customers are assertive and proactive enough to ask at the food service counter whether a product contains a food allergen, the customer may not get an appropriate or correct answer. Quick-service restaurants experience high staff turnover (NRA 2016a), making it more difficult (and providing less incentive) to train all staff on food allergen control practices. Insufficient training is likely one reason employees in limited-service restaurants have been demonstrated to lack knowledge of food allergen management (Dupuis et al. 2016). Quick-service workers may not realize that the amount of an agent that can trigger a reaction is much smaller for food allergies compared to food intolerances, nor appreciate that the consequences of accidental ingestion can be much more serious for those with food allergies than for those with food intolerances (Li 2007).

Information on food allergens may not be readily available at the restaurant, or staff may not have the knowledge to identify allergens from ingredient lists; for example, they may not recognize that "arachis oil" or "casein" could be problematic ingredients for customers allergic to peanuts or milk, respectively. Employees may not actually know if an allergen is present or not, or may not know how to find this information. In a busy quick-service restaurant, it might be easier for employees to incorrectly say "No" than to try to find out for sure if an allergen is present when a customer inquires about an allergen. Conversely, some restaurant staff may err on the side of caution and be unwilling (or unable) to ensure that a food is in fact allergen-free, restricting the ability of those with food allergies to order at restaurants (Leftwich et al. 2011).

In addition, quick-service servers "may not know what they don't know"; one survey found no correlation between a restaurant server's confidence in their own abilities to provide a safe meal to someone with a food allergy and their knowledge regarding food allergies (Ahuja and Sicherer 2007). Similarly, another study found that despite high confidence in being able to provide safe meals to those with food allergies, restaurant managers had clear deficits in their knowledge of and ability to manage food allergens (Wham and Sharma 2014). These risks and the related best practices on employee training to manage food allergens in restaurants are covered in more detail in this book (Chaps. 9 and 10).

11.3 Control of Food Allergens at Quick-Service Restaurants

11.3.1 Managing Supply Chain to Control Food Allergens

Quick-service restaurants require consistency across and efficiency within their establishments. As a result, quick-service restaurants often rely upon premade or processed foods obtained from central manufacturing facilities. Some raw ingredients such as produce, raw meats, raw fish, and eggs may be acquired locally by individual restaurants. Careful supply chain management for all of these ingredients is a major way in which quick-service restaurants can control food allergen risks. Supply chain management includes verification of supplier's allergen control programs, supplier approval and consistent use of only approved suppliers, a surveillance program, and corrective actions in response to supplier recalls, mock recalls, or customer complaints.

Ingredients that are used in preparing foods for a quick-service restaurant need first to be assessed for undeclared allergens. Quick-service restaurants need to ensure that each manufacturing facility maintains documentation for all ingredients used in each batch of product sold to the restaurant chain. Even minor ingredients or processing aids need to be scrutinized. In order to assess and manage risks of cross-contact, the quick-service restaurant should obtain from each food manufacturing facility supplying food ingredients and products a list of all allergens used in and by the manufacturing facility, and obtain documentation about how each are contained to prevent undeclared allergens. Verification of allergen prevention and cleaning programs using final product testing for major allergens used in the facility that are not declared in the product should also be utilized. The quick-service chain should require suppliers to retain samples of each batch of product if made on the same production line as a product that contains any of the eight major allergens in the event that later testing of a particular lot is necessary. New FDA requirements (in the form of allergen preventive controls) under the Food Safety Modernization Act (FSMA) require similar specifications to prevent undeclared allergens in manufactured human foods, and restaurants and chains can reinforce and monitor these preventive controls for the ingredients and foods they source in their supply chain (King and Bedale, 2017).

Each vendor of food ingredients for an independent quick-service restaurant or chain must be evaluated against a safety standard that includes allergen management to be approved by the quick-service restaurant before the vendor's products are used. This approval should be updated annually by the quick-service restaurant or chain.

All quick-service restaurants should use only approved brands and suppliers that have been evaluated properly for prevention of undeclared allergens. Individual quick-service restaurants must understand that they cannot make substitutions from non-approved vendors or brands (for example, if they run out of an ingredient, they should not use a similar ingredient from a different vendor unless it has been evaluated properly for potential presence of undeclared allergens). It may be a good idea to have approved backup suppliers in case the regular suppliers are not able to provide the needed ingredients. The quick-service restaurant should conduct self-assessments on a regular basis to ensure that only approved vendors and ingredients are being accepted from distributers and used to prepare foods. Bulk products received at individual restaurants (such as highly refined peanut oil used in frying) should be verified upon receipt to ensure that they came from an approved supplier.

Being part of a quick-service restaurant chain improves the efficiency of establishing approved suppliers. Importantly, being part of a large organization also makes it more efficient to monitor suppliers to ensure the safety of ingredients/products before they are distributed to restaurants in the chain. The parent company can institute surveillance for undeclared allergens in specific ingredients or products used by restaurant franchisees in a variety of ways. For example, they can routinely monitor related FDA warnings and product recalls and communicate this information to all restaurants when necessary. The parent company can also visit and inspect suppliers of ingredients and prepared foods or conduct product testing. For example, the parent company that specifies use of only peanut oil for frying could test the oil via a third-party accredited laboratory to ensure no peanut proteins are present as a way of ensuring that the oil from that supplier is highly refined and non-allergenic, and periodically reassess the product on a regular basis.

Establishment of a fast, effective method of implementing a product withdrawal or recall across all restaurants if an undeclared allergen is discovered through testing or reported by the FDA is also an important responsibility of the quick-service restaurant. All restaurants in even a very large quick-service restaurant chain could be contacted efficiently and rapidly through the use of interactive voice response systems (IVRS) and/or an e-mail/text system to alert each restaurant to the problem and to specify appropriate response actions. The system could also be used to automatically gather information on the response of each restaurant to the alert to ensure affected products are not sold to customers.

An individual quick-service restaurant and chain should also establish a system for customers to file complaints that might include food allergen problems, and the restaurant and/or chain needs to continually monitor complaints and promptly investigate them to ensure no ingredients or products have undeclared allergens. First, if one or more customers make a complaint about an undeclared allergen exposure or reaction to a product, and if that product is manufactured in a facility that also uses these undeclared allergens, then this product should be placed on hold for retail sales and further distribution to retail stores. If the product is found to contain an undeclared allergen (and this allergen is not used to prepare foods in the restaurant where the customer purchased the food), then a recall of that product based on production date should be communicated to remove that product from all restaurants and prevent any further distribution. Every customer complaint should be treated as valid until root cause assessments have been completed to confirm that there is no undeclared allergen in an associated product.

11.3.2 Retail Management of Allergen Risks: Restaurant Operations

The key steps in allergen control within a restaurant operation include food allergen training, food allergen management control programs (including fulfilling orders, preventing cross-contact), communication of food allergen information, corrective actions, and self-assessment.

Since 2005, the U.S. Food Code has included provisions to help restaurants protect consumers against food allergens (FDA 2005). These provisions include having a person in charge during operation who is knowledgeable about major food allergens, the symptoms of allergic reactions, and cleaning procedures needed to prevent cross-contact. That person is responsible for ensuring that other employees are trained in food allergen awareness as appropriate to their responsibilities at the restaurant.

Food allergen training is needed for both quick-service restaurant owners and employees. New quick-service restaurant owners should receive training on general food safety principles, including food allergens, prior to opening their restaurant and should be recertified on this training on a regular basis. When a new product is introduced, any potential food allergen risk of this product needs to be communicated to franchise owners as well as employees. Continuous awareness can be built through restaurant ownership and/or the parent company communications such as newsletters. Annual meetings are an ideal time to conduct allergen refresher training for franchise owners of a restaurant chain to both review important concepts on allergen risks and controls and demonstrate how important food allergen control is to the organization.

Challenges to training restaurant employees in a quick-service restaurant include high turnover, training costs, time for training, and language barriers (Kwon and Lee 2012). Food allergen training in a restaurant needs to be relevant, inexpensive, and accessible (Bailey et al. 2014). The National Restaurant Association has created an online food allergen training and exam program designed for restaurant employees and managers (NRA 2016b). The course, which is available in both English and Spanish, takes about 90 min to complete. Basic information on food allergies, symptoms, allergen identification, cross-contact prevention is presented, along with specific information for front of the house staff (communicating with customers, dealing with emergencies, etc.) and back of the house staff (reading food labels, handling food deliveries, etc.). Other local and state regulatory departments also offer food allergen safety training, such as the University of Minnesota Extension's *Food Allergen Training for Food Service Employees* course (see Chap. 12).

Food allergen training at a quick-service restaurant should be tailored to the employee's role within the restaurant and should include awareness of the eight major allergens and their presence in any menu items. Employees need to know the most common allergen questions a customer may have about the food served at the quick-service restaurant and be able to properly answer these questions to prevent miscommunication on the presence or absence of an allergen in a menu item. For example, one effective means to reduce risk of miscommunications on allergens to customers is to teach employees to *never* say the word "no" when asked if a product has a specific allergen; instead, offer the customer information that shows the ingredients of the product (e.g., nutrition brochure with ingredients of each product) so the customer can make an informed decision.

In addition to regular food allergen training, restaurant operators can implement programs to reinforce allergen control best practices among employees. All restaurant-specific food preparation recipes and job aid materials can be highlighted with visual cues to indicate when a procedure could introduce a food allergen risk (for example, the written materials describe precautions used when adding diced eggs to salads to prevent cross-contact of other products that share the same space). Continuous awareness of food allergies and the importance the company places in their control can be built through e-mails to employees, paycheck stuffers, and posters/job aids within the restaurant.

Ensuring allergen information is accurately transmitted from consumer to people taking orders and then to people preparing the foods is also important to consider in a quick-service restaurant. Methods of initiating and tracking special orders (for example, "no cheese" that might be a milk allergen concern) that are made by the customer should be established to alert staff in the kitchen to prepare foods properly and prevent foods from containing undeclared allergens.

Special tools, procedures, recipes, and specified food preparation areas can be used to help manage food allergen cross-contact (Fig. 11.1). Recipes can exclude the top eight allergens so that none of these allergens should be present in the restaurant prep kitchen, or the allergen (e.g., almonds) can be packaged by a supplier in a separate package (e.g., like a condiment) and then given to the customer to add to their product after purchase. Color-coded tools (e.g., purple for allergen-free zones only) can be provided that are only used in designated prep areas and can reduce the cross-contact of allergens to other foods.

Using disposable gloves is also an effective way to reduce cross-contact by food handlers working with foods containing allergens when special order request are made by customers; employees can remove gloves first, and then make the special order with a new pair of gloves. Note, however, that latex gloves themselves have been shown to trigger allergies when used to handle food eaten by consumers with latex allergies (Franklin and Pandolfo 1999; van Drooge et al. 2010), and some states do not allow latex gloves to be used in food service operations (Additives and

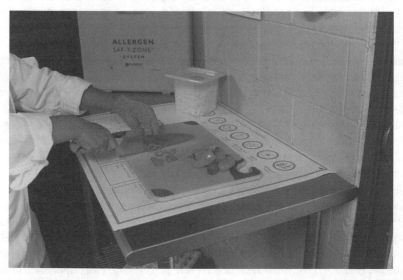

Fig. 11.1 Allergen-segregated food preparation space within a restaurant (photo used with permission from San Jamar). A separate food prep area and use of segregated utensils (only used for those products with the allergen(s)) can reduce the risk of cross-contact of allergens in other foods

Ingredients Subcommittee of the Food Advisory Committee 2003). Alternatives to latex such as nitrile and vinyl gloves are widely available and similarly priced.

Regular quality and food safety self-assessments of quick-service restaurants should include evaluation of food allergen controls to ensure Active Managerial Control of allergen risk (King 2016). Besides identifying potential problems at restaurants and establishing corrective actions for management of the hazard, the self-assessment process also serves to reinforce training for employees. Corrective actions when any behavior or action is noted that is not in accordance with the established standard operating procedure (SOP) for allergen management in the restaurant should include removal and discard of the potentially affected food product(s), retraining of employees, and/or notification to the person in charge to ensure the SOP is revised if needed.

11.3.3 Communicating Allergen Information to Consumers in a Quick-Service Restaurant

Even if customers are reluctant to make their food allergies known when dining at a quick-service restaurant, the restaurant needs to take steps to ensure customers know when an allergen might be present. Consumers with food allergies often plan ahead and will access a restaurant's website prior to visiting to identify how "food allergy-friendly" a restaurant is. Quick-serve restaurant chains, having a limited and relatively constant menu across many locations, are well positioned to use their website to inform customers of the allergen content of their products. Current technology can be used creatively and effectively to allow customers to access allergen information in real-time and also at home when planning to dine out. Ingredient information for any product can be made available on demand at the point of service (on restaurant signage), via the company website, or on a mobile app available on smart phones. For example, a mobile app for a quick-service restaurant business can make it easy and fast for consumers to determine whether or not a particular menu item contains a specific allergen (Fig. 11.2).

The point of purchase (POP) register and software systems used for ordering and tracking orders at a quick-service restaurant can be taken advantage of to generate special labels that can be used on special orders as they are being fulfilled, and the presence of those labels on the final product can alert the customer that they have indeed received the correct product. Other examples of how the technology infrastructure already present in a quick-service restaurant can be exploited to protect food allergic individuals exist. For example, if a customer inquires about an allergy they have at the time of ordering, a print-out that automatically includes a list of a product's ingredients with any food allergens highlighted (Fig. 11.3) can be generated, making it easy and fast for a consumer to know with confidence whether the food they are about to order contains a declared allergen. Undeclared allergens

Fig. 11.2 Mobile website
with allergen information
(Photo produced by Josh
King)

(not an ingredient in the product but from other ingredients used to prepare foods in
the kitchen) are, of course, also a concern to the customer that requires other types
of controls discussed above.

Food in restaurants is generally not served with a label or other notification that
could warn a customer that a potential allergen is present. However, unlike food in
full-service restaurants, quick-service restaurant food is often packaged in a wrap-
per or container, which represents an opportunity for the restaurant to notify customers that an allergen may be in the product. As an example, Fig. 11.4 provides an
example of avoidance messaging by using a simple statement on a fried chicken box
to alert customers to the presence of an ingredient that might be of concern to those
with a specific food allergy.

As an aside, while the FDA and other scientists have agreed that highly refined
oils such as peanut oil should not elicit allergenic reactions (Crevel et al. 2000) and
therefore do not have to be labeled as food allergens (FDA 2004), some customers
with severe peanut allergies habitually avoid all foods cooked in any type of peanut
oil. The inclusion of a label such as the one depicted in Fig. 11.4 (even if for a
refined peanut oil) clearly communicates that the product was cooked in peanut oil
and that it may contain peanut allergens, allowing the consumer to make an informed
decision whether or not to consume the food.

Fig. 11.3 Example of
customer print-out with
ingredient list

5569 Chicken Feet Drive
Peckingville, Georgia 00001
****

DATE:	02/02/2014
TIME:	11:11
ALLERGEN:	MILK
TYPE:	Cheese

CUSTOMER
NAME: ELSA B
SERVER: 34 ROSEBUD

INGREDIENTS

Water, Carrots, Onions, Red Lentils
(4.5%) Potatoes, Cauliflower, Leeks,
Peas, Cornflower, **Wheat** flour,
Cheese (**Milk**),Yeast Extract, Concentrated
Tomato Paste, Garlic, Sugar, **Soy**
Seed, Sunflower Oil, Herb and Spice,
White Pepper, Parsley.

ALLERGY ADVICE

For allergens, see ingredients in **bold**

Customer Copy

Fig. 11.4 Quick-service
packaging including
precautionary labeling

11.4   The Future

Technology is already being used to communicate risks within restaurants, train restaurant employees, and provide customers with real-time food allergen information. Technology will likely play an increasing role in communicating food allergen risks in the future.

In addition to developing ways to prevent food allergen problems via the methods described above, quick-service restaurants could establish the ability to delay and/or prevent anaphylactic reactions by medical means that might occur within their facility when exposure occurs. When prevention fails, prompt treatment of anaphylactic reactions with epinephrine can save lives. Since reactions often occur within 30 minutes of allergen ingestion (Jarvinen 2011), they may become evident before an individual leaves the restaurant. Unfortunately, many of those with food allergies do not always carry epinephrine autoinjectors with them; one survey in pediatric patients found that only 29% of those with a prior food allergic reaction requiring epinephrine were carrying (or had an accompanying family member present who was carrying) self-injectable epinephrine at the time of the survey (Curtis et al. 2014). The ability of restaurants to deliver emergency epinephrine injections to customers undergoing an anaphylactic reaction faces legal obstacles that many states are now trying to overcome, similar to initiatives in recent years to make epinephrine available in schools. For example, the restaurant (rather than an individual) must be allowed to obtain an epinephrine prescription not designated for use in a specific person. Staff must be trained in the proper storage and administration of the drug and be legally permitted to administer it. Liability considerations represent another potential hurdle. Currently at least 27 states have passed legislation that will make it easier for public entities such as restaurants to maintain and use epinephrine (FARE 2015). The availability of epinephrine at restaurants could be considered analogous to the presence of automated external defibrillators now found in many public facilities. Nevertheless, it remains important to train employees on how to recognize possible allergic reactions by customers and to initiate a call to 911 immediately.

The increasing prevalence of food allergies (Branum and Lukacs 2009) will improve awareness to them. More and more individuals working at quick-service restaurants will have direct experience with food allergies, making them perhaps the ideal ambassadors within their organization to teach others.

11.5   Summary and Conclusions

Quick-service restaurants face different challenges in dealing with food allergen risks than do full-service restaurants. Some, but not all individuals with food allergies, report having more confidence eating at large chain restaurants, including quick-service restaurants, than at independent restaurants (Kwon and Lee 2012). Controlling the supply chain and ingredients, managing restaurant operations (including employee training), and communicating with customers are key ways in which a restaurant chain can prevent food allergy problems. Quick-service restaurants must also overcome the fear of liability: Mistakes happen; negligence is the real problem. Finally, promoting a culture within the restaurant that recognizes the significance of food allergies and makes controlling food allergens a priority is an important responsibility for all retail food service businesses.

Acknowledgements The authors would like to thank Tong-Jen Fu, Lauren S. Jackson, and Kathiravan Krishnamurthy for coordinating the writing of this book and for organizing the conference on which this book is based. Hal King would like to thank S. Truett Cathy for allowing him to study and implement allergen control programs at Chick-fil-A Inc. quick-service restaurants.

References

Additives and Ingredients Subcommittee of the Food Advisory Committee. 2003. Food-mediated latex allergy: Executive summary. FDA Center for Food Safety and Applied Nutrition. Available at: http://www.fda.gov/ohrms/dockets/ac/03/minutes/3977m1.pdf. Accessed 22 August 2012.

Ahuja, R., and S.H. Sicherer. 2007. Food-allergy management from the perspective of restaurant and food establishment personnel. *Annals of Allergy Asthma and Immunology.* 98: 344–348.

Bailey, S., T. Billmeier Kindratt, H. Smith, and D. Reading. 2014. Food allergy training event for restaurant staff; a pilot evaluation. *Clinical and Translational Allergy.* 4: 26.

Banerji, A., S.A. Rudders, B. Corel, A.M. Garth, S. Clark, and C.A. Camargo. 2010. Repeat epinephrine treatments for food-related allergic reactions that present to the emergency department. *Allergy and Asthma Proceedings.* 31: 308–316.

Branum, A.M., and S.L. Lukacs. 2009. Food allergy among children in the United States. *Pediatrics.* 124: 1549–1555.

Crevel, R.W.R., M.A.T. Kerkhoff, and M.M.G. Koning. 2000. Allergenicity of refined vegetable oils. *Food and Chemical Toxicology.* 38: 385–393.

Curtis, C., D. Stukus, and R. Scherzer. 2014. Epinephrine preparedness in pediatric patients with food allergy: An ideal time for change. *Annals of Allergy Asthma and Immunology.* 112: 560–562.

Dupuis, R., Z. Meisel, D. Grande, E. Strupp, S. Kounaves, A. Graves, R. Frasso, and C.C. Cannuscio. 2016. Food allergy management among restaurant workers in a large US city. *Food Control.* 63: 147–157.

Eigenmann, P.A., and S.A. Zamora. 2002. An internet-based survey on the circumstances of food-induced reactions following the diagnosis of IgE-mediated food allergy. *Allergy.* 57: 449–453.

FARE (Food Allergy Research and Education). 2015. Public access to epinephrine. Available at: http://www.foodallergy.org/advocacy/advocacy-priorities/access-to-epinephrine/public-access-to-epinephrine. Accessed 22 March 2016.

Franklin, W., and J. Pandolfo. 1999. Latex as a food allergen. *New England Journal of Medicine.* 341: 1858–1858.

Jarvinen, K.M. 2011. Food-induced anaphylaxis. *Current Opinion in Allergy and Clinical Immunology.* 11: 255–261.

King, H. 2016. Implementing active managerial control principles in a retail food business. *Food Safety Magazine.* February/March.

King, H., and W. Bedale. 2017. Hazard analysis and risk-based preventive controls; improving food safety in human food manufacturing for food businesses. Elsevier.

Kwon, J., and Y.M. Lee. 2012. Exploration of past experiences, attitudes and preventive behaviors of consumers with food allergies about dining out: A focus group study. *Food Protection Trends.* 32: 736–746.

Leftwich, J., J. Barnett, K. Muncer, R. Shepherd, M.M. Raats, M.H. Gowland, and J.S. Lucas. 2011. The challenges for nut-allergic consumers of eating out. *Clinical and Experimental Allergy.* 41: 243–249.

Li, J. 2007. Mayo Clinic office visit: Food intolerance vs. food allergy. An interview with James Li, M.D., Ph.D. *Mayo Clinic Women's Healthsource* 11: 6.

NRA (National Restaurant Association). 2016a. Employee turnover rate tops 70% in 2015. Available at: http://www.restaurant.org/News-Research/News/Employee-turnover-rate-tops-70-in-2015. Accessed 25 March 2016.

———. 2016b. ServSafe allergen training. Available at: http://www.servsafe.com/allergens/the-course. Accessed 25 March 2016.

Rudders, S.A., A. Banerji, M.F. VassalloF, S. Clark, and C.A. Camargo. 2010. Trends in pediatric emergency department visits for food-induced anaphylaxis. *Journal of Allergy and Clinical Immunology.* 126: 385–388.

FDA (Food and Drug Administration). 2004. Food Allergen Labeling and Consumer Protection Act of 2004 (Public Law 108-282, Title II).

———. 2005. Food Code 2005. Available at: http://www.fda.gov/Food/GuidanceRegulation/RetailFoodProtection/FoodCode/ucm2016793.htm. Accessed 25 March 2015.

USDA-ERS (United States Department of Agriculture—Economic Research Service). 2015. Trends in U.S. food expenditures, 1953-2013. Available at: http://www.ers.usda.gov/data-products/food-expenditures/interactive-chart-food-expenditures.aspx. Accessed 25 March 2016.

———. 2016. Food Expenditures, Table 15. Available at: http://www.ers.usda.gov/data-products/food-expenditures.aspx. Accessed 25 March 2016.

van Drooge, A.M., A.C. Knulst, H. de Groot, C.J.W. van Ginkel, and S. Pasmans. 2010. Pseudo-food allergy caused by carry-over of latex proteins from gloves to food: Need for prevention? *Allergy.* 65: 532–533.

Weiss, C., and A. Munoz-Furlong. 2008. Fatal food allergy reactions in restaurants and food-service establishments: Strategies for prevention. *Food Protection Trends.* 28: 657–661.

Wham, C.A., and K.M. Sharma. 2014. Knowledge of cafe and restaurant managers to provide a safe meal to food allergic consumers. *Nutrition and Dietetics.* 71: 265–269.

Chapter 12
Food Allergen Online Training: An Example of Extension's Educational Role

Suzanne Driessen and Katherine Brandt

12.1 Introduction

The University of Minnesota Extension uses and enriches the university's research base by generating educational programming that provides value to society and in turn, back to the university (University of Minnesota Extension 2011). The University of Minnesota Extension food safety educators provide education to consumers and the food industry (including small food manufacturing facilities, food service industry, and retail operators) to support healthy, sustainable, and safe food at home and away.

The University of Minnesota Extension food safety educators have more than 15 years of experience developing and teaching food safety courses to the food service industry, training approximately 1000 participants annually. The University of Minnesota Extension team is known for their renewal online food safety course, reaching an average of 350 certified food managers (CFM) annually. In 2014, 6406 CFMs in Minnesota renewed their certificate with 378 or 6% meeting the training requirement through the Extension's online renewal course.

The state of Minnesota's Departments of Health and Agriculture have established an advisory committee composed of individuals from government, community organizations, and industry to periodically review proposed changes to the state's Food Code (Minnesota Department of Health). On January 25, 2011 the Minnesota Food Code Rule Revision Committee voted to adopt the Food and Drug Administration (FDA)'s 2009 model Food Code recommendations (FDA 2009, Minnesota Department of Health 2011) to require food allergy awareness training for food service employees. In response to this requirement and with encouragement from the Food Allergen Subcommittee of the Minnesota Food Code Rule

S. Driessen • K. Brandt (✉)
Extension, University of Minnesota, St. Paul, MN 55108, USA
e-mail: driessen@umn.edu; brand030@umn.edu

© Springer International Publishing AG 2018
T.-J. Fu et al. (eds.), *Food Allergens*, Food Microbiology and Food Safety,
DOI 10.1007/978-3-319-66586-3_12

Revision Committee, the University of Minnesota Extension food safety educators developed a web-based food allergen training program for food service employees. The creation of this online course is consistent with the food safety educators over-arching mission: to sustain the high quality food safety programs that currently exist and to be responsive to the need for new programming that supports a supply of safe food throughout Minnesota. In addition, reaching more participants via technology is a goal of Extension's strategic plan.

The *Food Allergen Training for Food Service Employees* course is a stand-alone, asynchronous, web-based course developed by University of Minnesota Extension food safety educators to provide essential information to food service employees regarding safe food handling for persons with food allergies. The course provides 60 min of interactive instruction via the University of Minnesota content manage-ment platform, Moodle, and is available 24/7 with self-access via individual or employer enrollment through the University's continuing education registration system. The registration fee of $25 can be paid online via the course link, after which time registrants have 3 months to complete the course.

The goal of the *Food Allergen Training for Food Service Employees* course is to increase safe food handling practices in food service settings, thus reducing the incidence of food allergic reactions. The course was developed to meet the food allergy awareness training requirement that the Minnesota Food Code Rule Revision Committee recommended for food handlers. The course provides current research, regulations and best practices in a learner-centered and interactive format.

The course content was designed to enable participants to do the following:

- Explain the difference between a food intolerance and a food allergy.
- Understand the seriousness of food allergies and recognize major food allergens.
- Describe their role to prevent a food allergic reaction.
- Explain the consequences of unsafe food handling practices related to food allergens.
- Analyze menu items for the eight foods that cause 90% of the food allergic reac-tions in the U.S.
- Recognize common symptoms of a food allergic reaction.
- Differentiate between cross-contact and cross-contamination control measures.
- Practice safe food handling practices to prevent a food allergic reaction.
- Describe the appropriate response to a food allergic incident.

In developing the course, andragogy principles with a focus on innovative tech-nologies were used to engage the learner and facilitate understanding. The course content and delivery are grounded in adult learning theories and incorporate key principles of health education behavior change theories. Hodell's (2000, 2011) instructional design process was used to develop an analysis-objective-driven cur-riculum. The goal is to deliver the best learner-centered experience possible. Engaging graphics, cause and effect stories, videos, multiple interactive activities, and narrated content were used to appeal to the varied learning styles of food service employees.

In this chapter, the *Food Allergen Training for Food Service Employees* course is showcased as an example of Extension's outreach and educational role. An outline of this ready-to-use training tool designed for the food service industry is highlighted. Literature findings that explain how food employees learn and describe what food employees do not know about food allergens support the background of the instructional design approach. A brief overview of the educational approach used is included, which applied adult learning principles to build understanding of the online course design process. Finally, the value and impact of the course are summarized, demonstrating knowledge gains and behavior change of the learner.

12.2 Literature Review

The project began by conducting a literature review to explore gaps in knowledge and food handling practices related to food allergens. In addition, it was critical to understand what was known about how food service employees learn and to identify effective educational approaches for the target audience. The findings were applied to the course design and are described in the following sections.

12.2.1 Literature Review Findings: Food Employees' Knowledge and Training Deficits

Researchers at Jaffe Food Allergy Institute (Bailey et al. 2011) found retail food employees had knowledge deficits about food allergens that could be harmful to diners. The study of 12 managers and 74 employees found:

- 25% believed "an individual experiencing a reaction should drink water to dilute the allergen."
- 23% said "consuming a small amount of an allergen is safe."
- 21% thought "removal of allergen from a finished meal would make it safe."

The Jaffe study and another (Ajala et al. 2010) demonstrated a lack of knowledge and considerable misinformation among food employees in regards to food allergens.

In addition to knowledge deficits, there are also training deficits. Minnesota Certified Food Managers (CFMs) and Persons-in-Charge (PIC) are required to provide food safety training to staff (Minnesota Department of Health 2014). In some cases, the CFMs and PICs do not have the skills or interest in training others. As a result, they do not provide the required training. To understand the need for an outside training provider, University of Minnesota Extension food safety educators conducted a survey in the fall of 2010 asking 73 CFMs about training practices for their employees and preferences for training. Results of the survey (Brandt and

Driessen 2010), which was not specific for food allergen training, verified the need and interest in an online food safety training program for food service employees:

- CFMs think they do an "okay" (22%, $n = 16$) to "fair" job (34%, $n = 25$) of training employees on policies, procedures and food handling practices to prevent foodborne illness.
- 34% ($n = 25$) would "definitely use" an online course to train employees.
- 21% ($n = 15$) like having a consistent provider (Extension) deliver the course.
- 23% ($n = 17$) had not taken Extension's online renewal course because they did not know about it, but now plan to take it in the future, illustrating how crucial marketing these programs are to their success.

12.2.2 Literature Review Findings: Food Employees' Learning Preferences

The literature was also reviewed to better understand how food employees learn. Beegle (2004) concluded that most food employees are oral culture learners. Oral culture learners need compelling messages in order for a food safety practice to be carried out 100% of the time. These messages could target prevention of a foodborne illness or a food allergic reaction incident. Beegle recommended communicating food safety concepts to oral culture learners by use of stories and sayings with vivid examples to allow the employee to "feel" the impact of a behavior. They need to know the "why" and "how" a food safety practice relates to the food they prepare, serve, and sell, which may challenge previous experiences and perceptions.

Tart (2012) described the difference between print and oral culture learners. Print culture learners seek out written information and need details and facts (Tart 2012). Most regulators and educators are print culture learners. Tart also noted that most food safety educational materials and teaching methods are developed by and for print culture learners. Print messages are often misunderstood or have little meaning to oral culture learners. Research for the Food and Drug Administration's oral culture learner project (Tart and Pittman 2010) tested educational materials such as storyboards, cause and effect posters, videos, and demonstrations, deeming these modalities as effective communication tools for food employees.

Food safety rules and regulations are fairly prescriptive, so why does not everyone understand them? A lack of comprehension may be related to a preferred learning style—how one receives and processes information that impacts understanding and motivation. Written and detailed food safety rules and regulations make sense to print culture learners like regulators and food safety educators but not to most food service employees with a preferred learning style that is oral in nature. Oral culture learners trust and seek out information from people they know. They need to connect personally with the information to recognize their behaviors may have positive or negative consequences. This personal connection can have a profound impact on employee's commitment to food safety and to protecting public health.

12.2.3   Literature Review Findings: Food Employees' Self-Efficacy and Cultural Background

In addition to knowledge of preferred learning styles, understanding the organizational culture and underlying health and cultural beliefs of the target audience are important dimensions educators and course developers should consider and understand before developing educational materials (Powell et al. 2011; Stuart and Achterberg 1997).

Knowledge of food safety rules and regulations is important, but it is just one piece of the equation. It is the practice of food safety behaviors that prevents foodborne illness outbreaks and food allergic reactions. Research by Schafer et al. (1993) used the health belief model to examine food safety practices. They found that understanding the threat of foodborne illness is just one way to motivate safe food handling behaviors. Food employees must also have self-efficacy (believe they are capable of performing a task) and have self-confidence that as an individual they can do something about food safety: in this case, prevent a food allergic reaction.

Differences in culture may impact health-related behaviors, as can differences in health-related beliefs (Kreuter et al. 2003). For instance, a traditional Hmong belief regards illness as the manifestation of the loss of spirits or souls (Cha 2003; Pinson-Perez et al. 2005). Preventing illness may be a bigger motivator towards adopting good food safety behaviors for this group than other motivators such as preventing financial or legal repercussions to a restaurant. Cultural background not only influences behaviors and beliefs, but it can also directly influence the level of acceptance of a health promotion program like food safety (Kreuter et al. 2003). Training techniques may need to be adjusted based on cultural beliefs to change perceptions and increase compliance.

12.3   Food Allergen Training for Food Service Employees Course Overview

Literature review findings identified deficits in food allergen knowledge and food handling practices of food service employees. The literature review also found that organizational cultures and employee self-empowerment influence food safety behaviors. These findings, along with the lack of readily available food allergen training programs that could satisfy Minnesota's requirement for food allergen training for food service employees, were gaps that extension educators sought to fill via a learner-centered approach in their web-based *Food Allergen Training for Food Service Employees* course.

The *Food Allergen Training for Food Service Employees* course is designed to educate food workers, food managers and regulatory and training professionals. The training targets a general audience working in food facilities including restaurants, catering establishments, school food services, daycares, community centers, churches, hospitals, health care facilities, food pantries, grocery stores, food markets, cooperatives, bakeries, convenience stores, and other entities that prepare or sell food to the public. The key course modules (Understanding Food Allergens, Managing Food Allergens, and Emergency Response to Allergy Reactions) are applicable to various food service roles including person-in-charge, front-of-the-house and back-of-the house duties. Waitstaff, cooks, and managers should complete the entire course to understand the important role that each person has in responding to a food allergen request.

The course can serve as initial training for those with limited or no knowledge of the subject or can function as a "booster/refresher course" for those with prior knowledge to provide consistency of food allergen information amongst all food workers and managers. The course can be taken to fulfill continuing education requirements for Minnesota certified food managers. Finally, the course serves as a professional development training opportunity for new food and health inspectors and as continuing education for registered sanitarians and food safety trainers.

As discussed earlier, research indicates most food employees learn orally (Beegle 2004; Tart and Pittman 2010). Oral culture learners require compelling messages to understand why following proper food safety practices are important. Therefore, an innovative storytelling approach was selected and woven throughout the course to appeal to these types of learners (Fig. 12.1).

In the course, learners hear Joe's food allergen story. Joe is allergic to peanuts, milk, and shrimp. He travels for his job, so eating out is a necessity. Joe counts on employees at grocery stores, delis, cafes and restaurants to provide safe food and prevent an allergic reaction. Joe's story is used to provide examples of how food

Fig. 12.1 Course screenshot featuring Joe's reaction

allergy reactions might occur in a food service establishment (such as when Joe touches a menu that had not been wiped off after milk spilled on it). His first-person narrative helps the learner empathize with the food allergic consumer and understand the serious implications of a reaction. The story evolves throughout the course so that in the Emergency Response module, Joe is hospitalized because of a mistake made by a food employee. The narrative uses this mistake (allergen cross-contact through the prior use of a serving spoon with an allergen-containing dish) to introduce the concept of cross-contact. Joe's story also demonstrates best practices for responding to a food allergy reaction, including calling 911 immediately, ensuring that the customer having the reaction remains seated, and later investigating the circumstances surrounding the reaction with all staff to identify and correct any practices that contributed to the reaction.

12.4   Instructional Design Course Development Process

Instructional design is the practice of creating "instructional experiences which make the acquisition of knowledge and skills more efficient, effective and appealing" (Hodell 2000). The process consists broadly of determining the needs of the learner, defining the end goal of instruction, and creating "interventions" to allow the learner to meet the goal.

Many instructional design models have been developed to provide framework and templates when developing instructional products. ADDIE is the acronym for a five-step course design process that provides designers and facilitators the structure and the flexibility to develop analysis-objective-driven curriculum with the ultimate goal of delivering the best learner-centered experience (Hodell 2000, 2011; Peterson 2003; Molenda 2003). ADDIE was chosen to develop the *Food Allergen Training for Food Service Employees* course because it gave designers and content experts the structure and focus to develop and align learner outcomes with content, activities and assessments in a user-friendly format. The ADDIE instructional design process involves performing each of five steps—Analysis, Design, Development, Implementation, and Evaluation—in sequential order (Ozdilek and Robeck 2009), with each step laying the foundation for the next. Working through each step is important to create a learner-centered course that prevents student problems and setbacks.

The *Food Allergen Training for Food Service Employees* course design utilized the collaborative adult model of learning (Stacey 2007). In this model, the learner is responsible for self-monitoring and self-assessment of skills. For example, the course starts with a short pretest on the user's preexisting knowledge of food allergen safety. The results of this test are presented to the user only at this point; the same test is later given as a posttest, allowing the user to see how much they have learned. The collaborative adult learning model was appropriate for the wide target audience because the procedural learning approaches built into the course can be used with a variety of learning styles and levels of food service experience.

Work-based assignments (such as identifying menu items that someone with a specific food allergy could order) were also included to help ensure training was relevant to the workplace.

The educators wrote, designed and developed the allergen training course as part of a 10-month University professional development workshop, *Learning Mostly Online*, funded by the Office of Distance Education and Technology. The project team included food safety educators as subject matter experts, the food safety team leader as project manager, a consultant from information technology, an instructional designer and programmer specializing in e-learning, and an advisory committee.

Each design phase along with examples from the development of the *Food Allergen Training for Food Service Employees* course are described in the following sections.

12.4.1 Analysis

The first phase of content development is to analyze and gather information about the target audience and to begin to focus on the project's overall goals while outlining "big picture" content.

A curriculum advisory committee which included external consultants was formed to analyze information from the literature review (discussed above in Sect. 12.2), to assess the target audience, and define the project's specific goals (listed in Sect. 12.1).

The target audience was identified as food service employees (managers, wait-staff, and cooks) in all retail food service facilities including restaurants, schools, day cares, health care facilities, and convenience stores. The main goal of the *Food Allergen Training for Food Service Employees* course was to increase safe food handling practices in food service settings, thereby decreasing food allergic reactions. The course needed to provide current research, regulations and best practices in a learner-centered online format. The course was designed for all target audience members to complete the entire course to emphasize the important role each person has in responding to a food allergen request.

Given the target audience, an online 1-h course was identified as a convenient, efficient and cost-effective method of training. The U.S. Department of Labor, Bureau of Labor Statistics (BLS) reports a >50% turnover rate for employees in the food service industry (NRA 2015). This high turnover rate suggests a need for ongoing accessible and affordable training.

There are many benefits of online courses for the food service industry, as discussed by Lee and Kwon (2011):

- Cost-effective for the food service business; Bell Canada saved $632 per attendee by switching to online training versus classroom training.
- Consistent and uniform information.

- Easier to update and keep current online than a written course.
- Self-paced.
- Available 24/7.

12.4.2   Design

The second phase of content development is design. In this phase, the developer outlines information gathered from the analysis phase. Educational theories and models of instructional design best suited for the target audience are selected. The project team begins to "drill down" and define the course content in detail. A sequence and flow for the course are established, with the overall project broken into manageable tasks.

During the design phase, the sequence of learning modules was defined as follows:

- Module 1: *Introduction*—presents an overview of course content and structure.
- Module 2: *Understanding Food Allergens*—provides background on food allergies, the foods associated with allergies, and symptoms of allergic reactions.
- Module 3: *Managing Food Allergens*—addresses how to respond to meal requests and prepare and serve food safely to prevent a reaction.
- Module 4: *Emergency Response to Food Allergy Reactions*—tackles response measures if an allergic reaction occurs.
- Module 5: *Summary*—offers a posttest, resources, evaluation, and certificate of completion.

Once the sequence of modules was established, the educators determined the learning objectives for the modules. The key question used to establish objectives was defining "what will the learner be able to do at the end of this course that they were not able to do before?" Learning objectives need to be consistent with the overall course-level objectives, be clearly specified from the students' perspective, and describe measurable outcomes. Nine learning objectives were identified and developed for the *Food Allergen Training for Food Service Employees* course, as presented in the bulleted list found in the Introduction of this chapter.

Once the learning objectives are clear, the next step is to develop strategies for the learner to achieve the desired objectives and determine the assessment strategy to measure whether or not the desired objectives are successfully achieved (Fig. 12.2). This is a critical process requiring much work to ensure objectives are aligned with content, activities and assessment. Students should be able to successfully complete the assessment activity if the objectives are embedded throughout the content, materials, resources, and learning activities.

Using a backward design approach (Wiggins and McTighe 2005) ensures the end product achieves the objectives. For the food allergen course, the educators started with the desired outcome and worked backwards to develop the other course components, i.e., content, activities, etc. For example, one desired outcome of the course

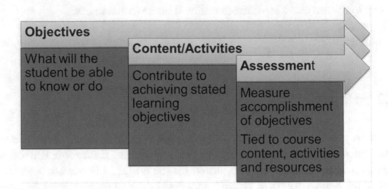

Fig. 12.2 Backward design process starting with objectives

was for participants to recognize common symptoms of a food allergic reaction. The content was then developed that would help achieve this outcome. In an early module, the participants are told that the symptoms of a food allergy reaction can range from a skin rash to difficulty breathing to death. Later in the module more detail is given about the variety of symptoms that can occur. Finally, a video clip with Joe speaking describes in his own words what it feels like to have a food allergy reaction. A short quiz is included in the module to test whether participants had retained the desired knowledge on food allergy reaction symptoms.

Another desired learner outcome in the *Understanding Food Allergens* module is for the learner to identify the "Big 8" foods that cause 90% of the food allergic reactions in the U.S. To deliver the content, the narrated presentation reviews the "Big 8" food allergens. The learning activity engages the learner by using a memory enhancer to remember the "Big 8." The assessment tool is a menu activity where the student decides what is safe for Joe to eat based on the content and activities thus far. Feedback is provided for both a right and a wrong response. Lastly, there is a workplace assignment to apply and reinforce the concept of reading labels for the "Big 8" food allergens. This example of content alignment is illustrated in Fig. 12.3.

During the design phase of course development, a detailed design document was created which outlined activities for the three project phases:

- Phase I: The Concept Proposal outlines background and context, specifies project purpose and goals, target audiences, identifies possible and very preliminary learner outcomes, and includes a general description of project deliverables.
- Phase II: General Project Planning profiles the target and secondary markets, provides specific details about deliverables and technologies available and needed for the project, describes staffing and support to develop the course, and discusses stakeholders and partners, intellectual property, funding and marketing.
- Phase III: The Instructional Design Plan shapes the design and development process, explores project staffing, project milestones and timeline, plans project

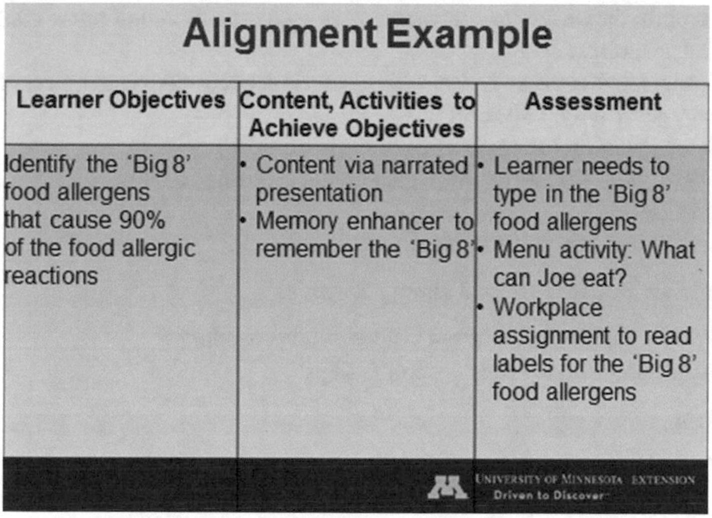

Fig. 12.3 Alignment of content with objectives

management, project evaluation, and quality assurance activities, develops an implementation and maintenance plan, and plans budget and other required resources.

The design process was aided by the development of a detailed timeline identifying the person responsible and a deadline for each task. Team members were required to commit to the timeline and carve out dedicated time to work on the project. The schedule included all dates from scripting to editing to programming to review to the final product. Highlighting dates when team members were not available helped to plan accordingly.

An outline of the final course content at the end of this phase is shown below:

1. Introduction to training program, materials, disclaimer
2. Understanding Food Allergens

   (a) What is a food allergy?
   (b) Know the difference between food allergy and food intolerances.
   (c) Identify the most common allergenic foods.
   (d) Identify the top eight food allergens and/or ingredients served on your menu.
   (e) Read labels for food allergens and hidden ingredients.
   (f) Identify the symptoms of an allergic reaction.
   (g) Understand the cause of food-induced anaphylaxis.

3. Food Allergen Response

   (a) Food service establishment policies, procedures, response plan.
   (b) Establishing clear communications with customers and food preparers.
   (c) Avoiding hidden food allergens via cross-contact.

(d) Explain the difference between cross-contamination and cross-contact control measures.
(e) Understand sources of cross-contact via shared utensils, equipment, cooking, garnishing, and serving.
(f) Identifying forgotten sources of cross-contact.
(g) Prevention measures of cross-contact during cooking, garnishing and serving.
(h) Corrective action if a mistake occurs.

4. Emergency response to food allergy reactions

   (a) Food service establishment policies and procedures
   (b) Appropriate emergency response steps

5. Summary

   (a) Celiac disease and gluten-free diets
   (b) Food allergen response: An opportunity
   (c) Summary
   (d) Online resources
   (e) Certificate of completion
   (f) Evaluation
   (g) References
   (h) Referral to our website, follow us on Twitter

12.4.3 Development

In the third phase, development shapes the course. Details of what is needed to produce the product are defined. This stage of course creation involves team meetings, hammering out the specific content, writing scripts, rewriting scripts, and creating the details of the learning and assessment activities to keep the course learner-focused.

During this stage, final scripts and activities were developed into storyboards. A storyboard is a helpful tool to capture each "screen" that will appear in an online course. It captures the script and the activity or assessment and illustrates what will appear on the screen. It also serves as a communication tool for team members and programmers. In addition, storyboards are a helpful check and balance system to make sure that content, activities, and assessment tools are aligned to meet learner outcomes.

When developing the storyboards, opportunities to reinforce information already presented or to introduce new material are often identified. In one module of the training, a video of Joe talking about a reaction he had in a Chinese restaurant is shown. While the story is ostensibly about Joe and his reaction, the training course also uses the opportunity to highlight several best practices that a restaurant should use for food allergic guests. For example, Joe asks that his stir-fry be prepared in its own wok. Joe also mentions that the restaurant manager called 911 even though Joe had given himself an epinephrine injection. The latter "best practice" is not explicitly mentioned within the training elsewhere but is learned within the context of the story.

The story of Joe continues with his son, Mike, who recently began working as a prep cook at a café. Joe says that Mike will never take a wild guess in answering a customer's question about an ingredient in a menu item; instead, he will ask the person in charge and the cook. The storytelling approach allows various viewpoints to be explored.

Various other best practices to prevent cross-contact are mentioned throughout the course in video, audio, or within Joe's story, including the following:

- Thoroughly clean surfaces, serving containers, and cooking equipment before cooking or plating food for food allergy customers (or use dedicated equipment).
- Check ingredient labels for possible hidden allergens.
- Use fresh, unused cooking oil, water, broth, or liquid when cooking for someone with a food allergy.
- Store allergen-free ingredients and garnishes away from other foods.
- Garnishes for those with food allergies should not be taken from food bars, buffets, or common service areas. Garnishing should be done with clean hands, gloves, or tongs.
- Extreme care should be taken to prevent spills or splashes when stirring or plating food to prevent sources of cross-contact.
- Wash hands before picking up an order to serve it to a customer. Carry it on a separate, clean tray or hand carry to deliver to the guest.
- Always verify the order with the guest when serving to prevent accidental mix-ups with another customer.
- Put takeout orders in a separate bag to avoid spills or contact with other food.
- If a mistake is made with an order, start over; do not try to remove an allergenic ingredient even if it just a garnish.

Figure 12.4 shows an example assessing how well the learner can apply what they have learned by asking them to choose what Joe can eat from a menu.

Clicking on each item will provide an explanation for any ingredient within the menu item that might make it unsafe for someone with a food allergy. This exercise allowed some of the dangers of hidden allergens to be highlighted. For example, the French fries may have been cooked in a fryer that was also used to cook shrimp, or the brownie might have nuts in it. Butter could be present in the rice or steamed broccoli that is served with the baked chicken breast.

The *Food Allergen Training for Food Service Employees* course utilized Articulate software programs—Engage, Quizmaker, and Presenter—to create modules and bundle audio recordings, videos, quizzes, and interactive activities into an automated format. This software helped create an engaging learner assessment activity.

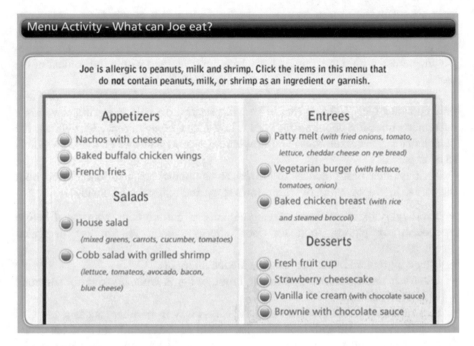

Fig. 12.4 Menu activity screenshot

12.4.4 Implementation

The fourth phase of creation of a training program using ADDIE is implementation. This phase includes usability and pilot testing (making adjustments as needed), finalizing all aspects of the course, establishing marketing strategies, and finally launching the course.

Usability and pilot testing were done by six advisory committee members and food safety educators. Reviewers evaluated course navigation, course content and course design. They were given clear instructions to review the questions before starting the course and tracked how long it took to complete the course. Suggestions and input were incorporated into the final product before launching the course.

Part of the implementation plan is marketing. The primary marketing source is the Extension food safety website (University of Minnesota Extension 2016). Potential customers can learn about the course, watch the promotional video and register for the course at the web site (http://www.extension.umn.edu/food/food-safety/courses/online/food-allergen-training/).

Multiple approaches continue to be utilized to reach the target audience, including Extension's Food Service training brochure (with distribution to 16,000+ CFMs and regulators), contact via e-mail to former students, Minnesota Environmental Health Association newsletters and meetings focused to regulatory partners

(Minnesota Department of Health, Minnesota Department of Agriculture, etc.), social media (http://twitter.com/umnfoodsafety) and a promotional video (https://www.youtube.com/watch?v=HSoFvKFOrm0).

12.4.5   Evaluation

The final phase is to evaluate the finished course to ensure the course achieves the desired goals. This final phase provides supporting evidence of success or indicates if alterations are needed.

Three types of post-course evaluation were incorporated in the course:

1. Formal testing of user knowledge (pretest and posttest) in order to certify course completion.

   A pretest and a posttest are included within the course. Because the same five questions are asked in both tests (although only the user can access pretest results), the pretests and posttests allow the user to assess how much they learned in the course. Asking the same questions twice also helps reinforce the content. A posttest score of at least 80% is required for certification of course completion. While these tests are used primarily to evaluate an individual's success in completing the course, the posttest results also provide some objective information on how successfully the course delivered the desired content. Examples of the questions asked include the following:

   • Food allergy and food intolerance are the same medical condition (True or False)
   • Touching a peanut butter cookie then a sugar cookie with the same glove can cause an allergic reaction in someone with a peanut allergy (True or False)

2. Quantitative user feedback on the course (online)

   Participants are directed to an online course feedback form upon completion of the *Food Allergen Training for Food Service Employees* course. The *Life Skills Evaluation System* (Washington State University Extension, http://ext.wsu.edu/LifeskillsNew/) was adapted to assess the learner's decision-making skills, use of resources, motivation, and self-responsibility to prevent a food allergic incident. *Life Skills* is a validated evaluation instrument for adult programming (Life skill evaluation system 2011). Evaluation questions measure growth in specific knowledge and skills as a result of taking the course. The instrument uses a retrospective pre-course/post-course question set. Bailey and Deen (2002) suggests this retrospective pre-course/post-course design minimizes bias related to self-reporting, eliminates pre-course and post-course coding, and is simple to implement.

   One of the goals for the course was to change food service worker's perceptions regarding food allergen management. Both Extension educators' experiences and literature findings indicated food service workers did not take food allergen management seriously and believed it is the customer's responsibility to prevent an allergic reaction.

The course feedback questions ask learners to rate their attitude regarding the seriousness of food allergens before and after taking the course. Importantly, the post-course evaluation found that 96% ($n = 101$) of participants greatly improved their appreciation of the seriousness of food allergens because of taking the course (Fig. 12.5). These findings are consistent with those of Choi and Rajagopal (2013), who found positive attitude changes related to the seriousness of food allergies increased with training, which their study also demonstrated was translated into safe handling practices.

The post-course evaluation indicated additional positive knowledge and behavior changes among the initial course participants ($n = 106$) including the following:

- Self-efficacy to implement safe food handling practices to prevent a food allergic reaction: before the course, 24% felt their ability to make safe food handling decisions to prevent an allergic reaction was "great"; this increased to 95% after taking the course.
- Ability to list the eight foods that cause 90% of the food allergic reactions in the U.S.: before the course, 48% said "none" to "slight"; at completion 94% were able to name the offending foods.
- Confidence to handle a food allergic emergency: before the course, 40% said "none" to "slight"; confidence level improved to 94% at completion.
- Looking at impact and behavior change, 87% of participants said that they would change their food handling practices to prevent a food allergic reaction at their foodservice establishment.

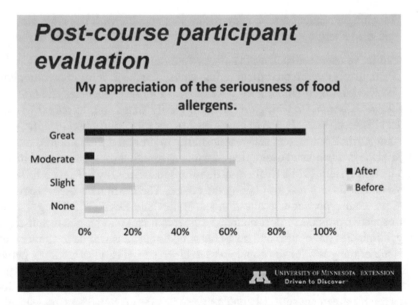

Fig. 12.5 Effect of course on appreciation of food allergen seriousness

3. Qualitative 3-month post-course evaluation (phone interview)

A 3-month post-course evaluation research study was conducted to assess knowledge retention, behavior change and system changes for individuals trained via the *Food Allergen Training for Food Service Employees* course. A literature review of evaluation techniques along with consultation with an Extension evaluation specialist determined a qualitative questionnaire via in-depth interview was the best instrument for the 3-month post-course follow-up evaluation study. This rigorous in-depth interviewing method requires fewer respondents (5–30), as more information can be collected by asking open-ended questions to gain in-depth understanding of the participant experience and identify themes representative of the population studied (Dworkin 2012). A quota selection sampling of 24 was utilized to include anyone who completed the course as tracked through the course content management system. Recruitment via e-mail to "opt in" or "opt out" recruited eight subjects (five of whom were CFMs) for the study.

The research study was conducted to address the following questions: (1) Did the course result in transfer of knowledge at the workplace to prevent food allergic reactions? (2) Did the course result in transfer of skills at the workplace to prevent food allergic reactions? (3) Did the course result in a policy or organizational change at food service establishments?

Questions included the following examples:

- Are online courses like this one a good way for employees to learn about food safety information?
- Since the course, have you observed improved food handling behaviors related to food allergens?
- Did the course result in a policy or some type of organizational change at your food establishment?

All of the CFM respondents felt that online courses were an effective way for employees to learn about food safety information. Sixty percent of participants implemented procedures to prevent food allergen cross-contact in their workplace following completion of the course. An example that was shared was having a designated peanut butter sandwich preparation area. Study results (Brandt and Driessen 2013) also indicated that 80% of the respondents planned to share or incorporate food allergen training within their food safety training programs. However, 40% still felt food allergens were not an issue in their food service operation, emphasizing the need for continued food allergen training in all food service establishments. This is a particular challenge when food allergen training is not incorporated into the food safety culture because a customer with a food allergy may walk into a food service operation at any time.

12.5   Additional Tips and Recommendations to Consider for Online Course Development

To design and develop an effective learner-driven online course requires careful planning, time, and resources. Consider the following recommendations for online course design:

- Time, resources and commitment from all stakeholders are key factors to design and implement an effective learner centered course. Synergism truly makes a better course.
- Recruit an advisory group of food service industry, regulatory, and Extension representatives to guide the course of action on content, provide feedback on course development, and field-test the course.
- The *Food Allergen Training for Food Service Employees* course was selected for a third party scholarly review by Quality Matters (QM), the current quality assurance "gold standard" for learner centered online education, and was certified as meeting Quality Matters review standards. The use of such a rubric throughout the development and review process (https://www.qualitymatters. org/rubric, University of Maryland) to evaluate all course components can be valuable in demonstrating to others the quality of your course (Quality Matters Rubric 2015).
- Take the time to complete an instructional design document. It is your course blueprint. Include your goals, target audience, course overview, course and module objectives, course outline, implementation plan, evaluation plan, staffing plan, and marketing plan for the project. This should be a flexible document as it will change and be updated as the project moves forward. Have a vision of what you want the course to look like so when you brainstorm with programmers, course designers, support staff, and others you can identify what is and is not possible. For example, Extension's course development team wanted to create their own video segments but soon found out it was not feasible due to staffing, budget and timeline constraints.
- Be realistic about the time required to fully develop a course. It was a 10-month process to develop the *Food Allergen Training for Food Service Employees* course. Dedicate the needed time to do it right.
- Evaluate each team member's strengths, skills, likes, and dislikes to determine roles and assignments. It will be a better product and process if team members' contributions align with their strengths.
- Communication is key! Deviations from the schedule can occur. Have a backup plan, ask for help, shift things around, and keep everyone informed about the changes.
- Designate someone who wants to be the project manager or outsource this role. It is critical to have a point person to keep the project on track and schedule. Build in time for changes and updates to the course after pilot and usability testing.

- Connect with Cooperative Extension in your state or county. Extension contributes to and draws upon University research to develop and deliver effective educational programs. It is a two-way process—Extension's educational model requires strong partnerships and networks. Successful Extension work provides value through education as indicated by our program outcomes and impacts.

References

Ajala, A.R., A.G. Cruz, J.A. Faria, E.H. Walter, D. Granato, and A.S. Sant. 2010. Food allergens: Knowledge and practices of food handlers in restaurants. *Food Control.* 21: 1318–1321.

Bailey, S., R. Albardiaz, A.J. Frew, and H. Smith. 2011. Restaurant staff's knowledge of anaphylaxis and dietary care of people with allergies. *Clinical and Experimental Allergy.* 41: 713–717.

Bailey, S.J., and M.Y. Deen. 2002. Development of a web-based evaluation system: A tool for measuring life skills in youth and family programs. *Family Relations.* 51: 138–147.

Beegle, D. 2004. Oregon environmental health specialist network communication study, Oregon Department of Human Services. Available at: https://public.health.oregon.gov/ HealthyEnvironments/FoodSafety/Documents/ehsnet.pdf. Accessed 4 June 2017.

Brandt, K., and S. Driessen. 2013. Food allergen training for food service employees evaluation study. Retrieved from the University of Minnesota Digital Conservancy. Available at: http:// purl.umn.edu/160486. Accessed 4 June 2017.

———. 2010. Employee and online training survey: What do the results mean? Implications for marketing and program planning. Unpublished report. Department of Extension. University of Minnesota, St. Paul, MN.

Cha, D. 2003. The Hmong 'Dab Pog Couple' story and its significance in arriving at an understanding of Hmong ritual. *Hmong Studies Journal.* 4: 1.

Choi, J.H., and L. Rajagopal. 2013. Food allergy knowledge, attitudes, practices, and training of foodservice workers at a university foodservice operation in the Midwestern United States. *Food Control.* 31: 474–481.

Dworkin, S.L. 2012. Sample size policy for qualitative studies using in-depth interviews. *Archives of Sexual Behavior* 41: 1319–1320.

FDA (Food and Drug Administration). 2009. Food Code 2009. Available at: http://www.fda.gov/ Food/GuidanceRegulation/RetailFoodProtection/FoodCode/ucm2019396.htm. Accessed 17 June 2016.

Hodell, C. 2000. *ISD from the ground up: A no-nonsense approach to instructional design.* Alexandria, VA: Association of Talent Development (formerly known as American Society for Training and Development).

———. 2011. *ISD from the ground up.* Alexandria, VA: American Society for Training and Development Press.

Kreuter, M.W., S.N. Lukwago, D.C. Bucholtz, E.M. Clark, and V. Sanders-Thompson. 2003. Achieving cultural appropriateness in health promotion programs: Targeted and tailored approaches. *Health Education and Behavior.* 30: 133–146.

Lee, Y. M., and J. Kwon. 2011. The effectiveness of web-based food allergy training among restaurant managers. Available at: http://scholarworks.umass.edu/cgi/viewcontent.cgi?article=12 36&context=gradconf_hospitality. Accessed 4 June 2017.

Life skill evaluation system. 2011. Life skill evaluation system. Washington State University. Available at: http://ext.wsu.edu/LifeskillsNew/. Accessed 4 June 2017.

Minnesota Department of Health. n.d.. Minnesota Food Code Rule Revision Advisory Committee. Available at: http://www.health.state.mn.us/divs/eh/food/code/2009revision/committees/. Accessed 17 June 2016.

————. 2011. Minnesota Food Code Rule Revision Advisory Committee meeting minutes. 01/25/2011. Available at: http://www.health.state.mn.us/divs/eh/food/code/2009revision/11_0125mtg/minutes.pdf. Accessed 21 June 2016.

————. 2014. Minnesota Certified Food Manager (CFM). Available at: http://www.anfponline. org/docs/default-source/legacy-docs/mn/documents/mn-certified-food-manager-fact-sheet. pdf?sfvrsn=0. Accessed 17 June 2016.

Molenda, M. 2003. In search of the elusive ADDIE model. *Performance Improvement.* 42: 34–37.

NRA (National Restaurant Association). 2015. Hospitality employee turnover rose in 2014. In News & Research. Available at: http://www.restaurant.org/News-Research/News/Hospitality-employee-turnover-rose-in-2014. Accessed 4 June 2017.

Ozdilek, Z., and E. Robeck. 2009. Operational priorities of instructional designers analyzed within the steps of the Addie instructional design model. *Procedia-Social and Behavioral Sciences.* 1: 2046–2050.

Peterson, C. 2003. Bringing ADDIE to life: Instructional design at its best. *Journal of Educational Multimedia and Hypermedia.* 12: 227–242.

Pinson-Perez, H., N. Moua, and M.A. Perez. 2005. Understanding satisfaction with Shamanic practices among the Hmong in rural California. *International Electronic Journal of Health Education.* 8: 18–23.

Powell, D.A., C.J. Jacob, and B.J. Chapman. 2011. Enhancing food safety culture to reduce rates of foodborne illness. *Food Control.* 22: 817–822.

Quality Matters Rubric. 2015. Quality Matters Rubric. University of Maryland. Online. Available at: https://www.qualitymatters.org/continuing-and-professional-education-rubric-program. Accessed 4 March 2017.

Schafer, R.B., E. Schafer, G.L. Bultena, and E.O. Hoiberg. 1993. Food safety: An application of the health belief model. *Journal of Nutrition Education.* 25: 17–24.

Stacey, E. 2007. Collaborative learning in an online environment. *International Journal of E-Learning and Distance Education.* 14: 14–33.

Stuart, T.H., and C. Achterberg. 1997. Education and communication strategies for different groups and settings. *FAO Food and Nutrition Paper*: 71–108.

Tart, A. 2012. Modifying the behavior of food employees. Using educational materials educational materials designed for oral culture learners. Presentation slides. U.S. Food and Drug Administration. Atlanta, GA. Available at: https://www.fsis.usda.gov/wps/wcm/connect/b96d8482-e3fd-470b-8426-f9facc3fe082/Slides_FSEC_ATart_Oral.pdf?MOD=AJPERES&CACHEID=60451509-cb4b-43f7-8283-b3ba45a531c0. Accessed 4 June 2017.

Tart A., and J. Pittman. 2010. Changing the behavior of food employees using materials designed for oral culture learners. Presentation. AFDO 114th Annual Education Conference, Norfolk, VA.

University of Minnesota Extension. 2016. Food allergen training for food service employees. Available at: http://www1.extension.umn.edu/food-safety/courses/online/food-allergen-training/. Accessed 22 February 2016.

————. 2011. Guidelines for decision-making: Extension strategic plan. Available at: http://www. extension.umn.edu/about/facts/extension-strategic-plan-2011.pdf. Accessed 17 June 2017.

Wiggins, G.P., and J. McTighe. 2005. *Understanding by design.* Alexandria, VA: Association for Supervision and Curriculum Development.

Chapter 13
Allergen Control at Home

Binaifer Bedford

13.1   Introduction

For many families in the U.S., discussions surrounding safety at home have evolved beyond babyproofing a residence or fire safety protocols. The increasing prevalence of food allergies in the U.S. has added strict allergen avoidance as a new dimension to the safety landscape for food allergic individuals and those who care for them (Gupta et al. 2011). The effort to minimize risk, encourage good health, and ensure the emotional well-being of the food-allergic family member(s) in a home environment manifests itself in many forms. Approaches to controlling allergens at home often vary depending on the allergic individual, family situation, and personal philosophy. This chapter delves into general factors to consider when developing a customized food allergen control plan best suited for the unique needs of each family.

While studies by Bock, Muñoz-Furlong, Xu, and others have documented that the majority (>65%) of fatal food-related anaphylactic reactions occur outside the home (Bock et al. 2001, 2007; Muñoz-Furlong and Weiss 2009; Xu et al. 2014), much can be learned from each situation. Some common factors that may have contributed to the tragedies include delayed administration or lack of available epinephrine, coexisting health conditions such as asthma, the age of the food allergic individual (primarily teens and young adults), and the type of food (cookie, candy, Asian food) or allergen (peanuts, tree nuts, milk, seafood) consumed. Educating food allergic individuals and those who care for them to promptly recognize the signs/symptoms of a reaction and to respond immediately by using an epinephrine auto-injector and calling 911 can result in a favorable outcome, especially for high-risk patients with asthma. Encouraging the food allergic consumer to seek allergen

B. Bedford (✉)
Division of Food Processing Science and Technology, U.S. Food and Drug Administration,
Bedford Park, IL, USA
e-mail: binaifer.bedford@fda.hhs.gov

© Springer International Publishing AG 2018
T.-J. Fu et al. (eds.), *Food Allergens*, Food Microbiology and Food Safety,
DOI 10.1007/978-3-319-66586-3_13

information on a product label, clearly communicate their food allergy, and assess a food establishment's ability to serve "safe" meals can help minimize the risk of an adverse reaction. Awareness of foods that are considered high risk for some allergic individuals based on product recalls is another factor to consider. For example, milk allergic consumers and their caregivers should be aware that bakery, chocolates/candy/confections, and dessert food categories are considered high risk for undeclared milk allergens (Gendel and Zhu 2013; Gendel et al. 2014). Similarly, the avoidance of some ethnic restaurants (i.e., Thai, Chinese, and Indian) that commonly use ingredients like peanuts and tree nuts in many recipes is also recommended unless there is a favorable record of a particular restaurant's ability to successfully serve food-allergic customers.

Understanding these risk factors and considerations can help one develop and strengthen an allergen control plan for the home. Ideally, the plan should include personal/family philosophy on the extent of allergen introduction into the home, product label review, identifying entry points, minimizing cross-contact situations and strategies for social events involving carryout restaurant food or pot-luck meals. Finally, the plan should include education/training for family members/visitors/caregivers on the symptoms of an allergic reaction, how to use an epinephrine auto-injector and the procedures for following the medical emergency plan.

13.2 Family Philosophy and Sibling Considerations

The development of a verbal or written home allergen control plan allows for early discussions surrounding the family philosophy with respect to food allergens and determining what is best for the affected individual(s) and other family members. Clearly defining the ground rules for the introduction or avoidance of specific allergens into the home sets the tone for the level of comfort or the degree of vigilance that must be maintained in the home environment. Factors that often need to be considered include the age and maturity level of all household members (food allergic individual(s), siblings, extended family living in the same residence), the number of allergens that must be avoided (single or multiple food allergens) and other special medical, dietary or nutritional needs within the immediate family. For example, a family with only one child having one food allergy may decide to allow the allergen into the home if they are able to provide appropriate controls to minimize accidental exposure. On the other hand, some families with multiple young children or those having to control for multiple allergens may choose to exclude the allergen(s) from the home environment, especially if there is low confidence in their ability to contain and limit the spread of the allergen(s) throughout the residence. In contrast, other families might find that excluding multiple allergens for all members of the family is too restrictive and challenging from a nutritional perspective, so they devise systems that allow and control allergens in the home. Again, it is important to remember that there is no right or wrong approach regarding the extent of food allergen introduction into a home, and each situation is unique.

Periodic reassessment of the allergen control plan allows for increased responsibility and participation of the food allergic member in the decision making process in an age-appropriate manner. A good time to re-evaluate allergen controls in the home is after the annual physician/allergist appointment when the patient's medical history is reviewed and reassessed. Likewise, incidents involving a serious allergic reaction can also modify the family philosophy and may result in changes that increase safeguards to the home allergen control plan. Ultimately, the decision to include or exclude allergens from a home is a personal/family choice that depends greatly on the specific situation and is typically modified based upon the past and current history of reactions.

13.3 Label Review and Entry Points

13.3.1 Reviewing Labels for Allergen Statements

Whether a family chooses to exclude the allergen(s) from the home or control the introduction of the allergen within the home environment requires reading labels. The Food Allergen Labeling and Consumer Protection Act (FALCPA) of 2004 went into effect in 2006 and required all FDA-regulated prepackaged foods to clearly disclose the presence of the eight major food allergens (peanuts, tree nuts, milk, egg, soybeans, wheat, fish, and crustacean shellfish) in food products sold in the U.S. on product labels using plain English when added intentionally as a food ingredient (FDA 2004). Details regarding FALCPA, allergen labeling requirements, and exemptions (highly refined oils) are explained in depth on the FDA website and in the Food Labeling Guide (FDA 2013). Additionally, it should be noted that the specific type of tree nut (e.g., walnuts, pistachio, almonds, or pecans) and species of fish (e.g., trout, cod, or salmon) or crustacean shellfish (e.g., crab, lobster, or shrimp) are required to be clearly declared on the ingredient statement.

Two acceptable formats for allergen declarations on product packaging are illustrated in Fig. 13.1 as shown below (FDA 2013).

It is also important for the consumer to understand that the use of precautionary or advisory labels is not regulated. Since FALCPA went into effect in 2006, there has been a proliferation of phrases on packaged products such as "may contain …," "processed in a facility with …," "made on shared equipment with …," "may contain traces of …," and many other similar statements, which leads to ambiguity regarding the presence of an allergen in a product and consequently, risk-taking behavior on the part of some consumers (Pieretti et al. 2009; Crotty and Taylor 2010; Ford et al. 2010; Hefle et al. 2007).

Products bearing an advisory or precautionary label statement pose a dilemma for some consumers since they might contain allergens at levels that can elicit an allergic reaction (Gendel and Zhu 2013; Pieretti et al. 2009; Simons et al. 2005). Due to this uncertainty and lack of regulation surrounding the use of these label

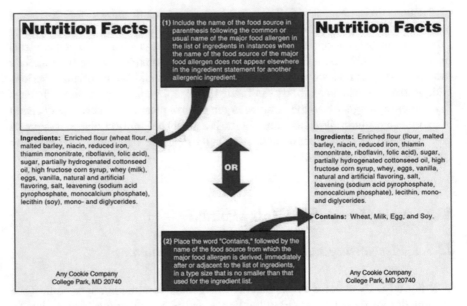

Fig. 13.1 Example of major food allergen declarations on labels for foods sold in the U.S. (Source: FDA Guidance for Industry: A Food Labeling Guide (FDA 2013))

statements, many food allergic consumers chose to purchase products that do not list the allergen on the ingredient statement or seek items that are produced in dedicated facilities that are allergy friendly or free from the allergen of concern.

Families who choose to avoid the introduction of the select allergen(s) into their home frequently decide to avoid purchasing food products with advisory claims in an attempt to keep their home a safe haven for the food allergic family member(s). While FALCPA applies to FDA-regulated products, the food allergen labeling requirement is not mandated under U.S. Department of Agriculture (USDA) Food Safety and Inspection Service (FSIS). However, the FSIS has issued guidance documents and best practice recommendations for meat and poultry products with respect to the control and declaration of allergens (USDA-FSIS 2015).

Similarly, it is also important for the consumer to understand that the Department of the Treasury's Alcohol and Tobacco Tax and Trade Bureau (TTB) has authority over most alcoholic beverages, and the labeling of allergenic ingredients per FALCPA is not mandated for TTB regulated products (FDA 2014a; TTB 2012; TTB 2014). The young adult consumer with food allergies should especially be aware of this fact and understand that many alcoholic beverages may not list allergens on the label. The exception to the TTB regulated products are the select FDA-regulated alcoholic beverages which include certain beers made without traditional beer ingredients (FDA 2014a). Consequently, the breadth and depth of label review typically extend beyond FDA-regulated food and beverage products to also include merchandise under the oversight of the USDA and TTB, with each agency issuing different allergen-related guidance documents.

Family discussions regarding label review, avoidance or acceptance of products with advisory labels, and understanding the potential risks regarding the presence of allergens in products with precautionary statements provide a foundation surrounding the extent of allergen controls that may need to be considered when developing an allergen control plan.

As simple as label review sounds, this task demands time, attention and concentration, especially as it is required for nearly every food item that is purchased. Consumers who are allergic to other foods beyond the eight major allergens that are regulated by the FDA must exercise additional vigilance when reading labels. Shopping with infants, toddlers, and young children can also add an element of distraction and result in quickly scanning a label and inadvertently purchasing a product with an allergen or advisory label. One strategy to help minimize such errors involves reading the ingredient statement multiple times, often reading the ingredient statement forward and then in reverse order for the allergen of interest. Recruiting another individual to double check the label statement is also helpful. As food-allergic children grow older, it is often useful to involve them in the label reading process to build "safe" habits/behaviors and to gradually shift the allergen management responsibility to them in age-appropriate ways.

13.3.2   Food/Beverage Allergen Entry Points

Entry points for the introduction of allergens into a home typically arise from shopping trips to the grocery store, carryout or delivery of restaurant food, pot-luck food items, treats brought home from school/work/sporting events, and scenarios involving celebrations or holidays.

The most obvious way to control allergen introduction or entry into a home is through label review of food items during grocery shopping. The accuracy of the manufacturer's product labels per FALCPA either builds trust and loyalty to a brand or can diminish it if the manufacturer is subject to frequent allergen-related recalls. Consumers who permit the entry of the allergen(s) into the home as the direct product (e.g., liquid milk, shell eggs, peanuts, and tree nuts) or in the form of products bearing a precautionary statement must carefully consider the implementation of cross-contact controls to avoid unintentional exposure to the food allergic individual.

Ordering carryout meals from restaurants presents a situation that is similar to dining out. A recent study by the Center for Disease Control (CDC) highlighted that important gaps exist regarding food allergy knowledge and attitudes among restaurant managers and staff (Radke et al. 2016). Ideally, it is best to purchase carryout food from restaurants that have successfully served allergy friendly meals. If possible, investigate the restaurant in person instead of inquiring by phone. Begin the discussion with the appropriate individual, usually a manager or chef, during non-peak meal times. Obtain information regarding the restaurant's ability and willingness to work with customers having special dietary needs. Ask questions regarding staff training on allergens, meal preparation, use of dedicated equipment (fryers, grills, waffle irons, cutting boards, etc.), how the kitchen staff avoids cross-contact, and similar prompts.

Excellent resources for dining out for individuals with food allergies are now available online and highlight best practices that can be used when placing carryout orders as well (FARE 2017a). Ensuring clear communication and establishing trust with the food service provider can minimize the entry of unintended allergens into the home and decrease the risk of an allergic reaction. Additionally, some food allergic families opt to avoid the delivery option with carryout food to further minimize the number of transfers involving individuals who handle the meal (from the restaurant staff to the delivery driver) and feel more comfortable when the meal is picked up directly from the controlled environment of the restaurant kitchen.

Other entry points for allergens into the home include pot-luck dishes prepared by friends in a home kitchen environment, a retail food service operation such as a grocery store deli, coffee shops or bakeries. The same approach with questions similar to those used for restaurants and carryout situations is also applicable to these scenarios.

13.3.3   Non-food Products: Hidden Allergen Entry Points

Non-food or beverage items that are intended for ingestion must also undergo label scrutiny for the presence of allergens as ingredients or advisory/precautionary statements. This includes products such as vitamins, dietary supplements, protein powders, or medicines.

Families that choose to avoid the introduction of specific allergens into the home often look at other potential sources of allergen entry including dermal contact. With the expansion of botanical and natural products on the market, it is now common to find a variety of food allergens in products that are intended for topical application. This may include items such as soaps, lotions/creams, lip balms, hair and nail care products, cosmetics, oral care products, household cleaners/polish, art/craft supplies, and even potting soil/compost (Kim 2011). Case studies in the literature have documented that some individuals can react to food allergens in cosmetics and soap (Glaspole et al. 2007; Laurière et al. 2006; Pootongkam and Nedorost 2013). Dermal allergic reactions have also been attributed to products such as milk in soaps/cleansers, hydrolyzed wheat protein in cosmetics/shampoo/conditioner, and tree nuts (e.g., almond or walnut) in exfoliant cleansers. Common complaints included exacerbated eczema or an allergic reaction involving hives/rash for some sensitive individuals. Consumers should also know that cosmetic products as well as their ingredients and labeling are not covered under FALCPA and do not require FDA approval (FDA 2017).

Another common entry point and source of inadvertent exposure to food allergens is pet food. Controls surrounding the handling of pet food or accidental ingestion of the product are of concern especially in homes having toddlers or young children.

Family philosophy regarding food allergens in pet food also needs to be considered beyond label review to determine if there are food alternatives for the pet that do not contain the allergens of concern. Awareness of this hidden source of allergens and the potential to transfer it to other areas of the home by the pet or young children is an important factor to consider when obtaining a pet or changing the pet's diet.

At the other end of the spectrum are families who choose to allow allergens in their home in various forms, but have safeguards in place to minimize risk and control the spread of the allergen. Label review still plays an important role in the decision to permit allergens into the home environment, but the introduction of allergens means an additional level of vigilance and hand-washing may be needed to minimize accidental exposure due to cross-contact. If an allergen-containing pet food is permitted in the home, avoiding the saliva and dermal contact when interacting with the family pet is essential for sensitive individuals. For example, peanut butter or fish (salmon) are a common ingredient in some dog treats and dog food and require extra caution in households that are controlling for peanut or fish allergens.

13.4   Cross-Contact Controls

Personal experiences and philosophy guide each individual family with respect to the level of cross-contact controls for allergen management within their home. Environments where allergen entry is restricted often have fewer controls within the household beyond label review and handwashing. For situations where allergens are allowed into the home, cross-contact controls typically begin in the kitchen. Factors that need to be considered include physical segregation of allergenic products during storage, transfer of label and allergen information if products are not kept in original packaging, timing/sequence for meal preparation (i.e., allergen-free meal first), use of dedicated utensils/tools/equipment, zoning/restriction regarding seating arrangement, and effective cleaning.

Special attention to details such as the organization of storage space is needed to avoid cross-contact. Allergen segregation within the refrigerator, freezer, or pantry where food will be stored is a common technique. Specific shelves, bins or containers for products containing the allergens are typically used and often kept out of reach of young children. Establishing a consistent labeling system that also includes the transfer of label and allergen information to secondary containers helps ensure that products are clearly identified as either allergen-free (i.e., safe for consumption) or containing the allergen. This step is critical in communicating and minimizing allergen risks.

Allergen cross-contact controls before meal preparation begin with thorough hand washing and ensuring that food preparation and cooking surfaces are clean. The sequence in food preparation is also important and when possible, it is desirable to cook the allergen-free meal first in order to reduce the chance of exposure to the allergenic food. During meal preparation, dedicated cutting boards, cookware, utensils/cutlery, dishware/china can be employed to minimize the potential of allergen transfer.

Some families also choose to have dedicated equipment for kitchen items that are difficult to clean, such as toasters and waffle irons, to avoid cross-contact with breads or batters that may include allergens such as wheat, milk, egg, or nuts. Fryers and shared cooking oils are also sources of potential allergen cross-contact (Lehrer et al. 2007; Lehrer et al. 2010). The use of dedicated fryers can minimize the transfer of allergens from one food product to another. Awareness of actions and behaviors during meal preparation is important to avoid simple cross-contact scenarios. For example, it is imperative to remember not to accidentally use the same knife to spread dairy butter or nut butters on bread and then insert the same knife into a jelly/jam container, thus unintentionally resulting in a cross-contact situation with the jam/jelly and various allergens (wheat, milk, peanut, or tree nuts) that are not on the product label. The same holds true for the dangers surrounding other shared utensils such as stirring a cup of coffee containing milk, then reusing it in another food product, thus increasing cross-contact risks.

Creating dedicated seating space in the kitchen or at the dining table as an allergen-safe zone is another option that some families find helpful as a way to control the spread of allergens within the home, especially when young children and siblings are involved. One mother shared her strategy to protect her 3-year-old son with multiple allergies (milk, egg, peanuts, and tree nuts) from accidental exposure that could arise from kitchen interactions with his 7-year-old and 2-year-old siblings. In addition to preparing allergen-free meals first and having designated color-coded cups, dishware, and seating arrangements, additional precautions were taken with the younger toddler by using cups with lids and straws to minimize milk spills. The allergic 3-year-old was educated on the dangers of using cups or utensils from others, lessons that were learned as a result of several accidental exposures and resulting emergency room visits from daycare mishaps. This family understood the dangers of sharing cups, utensils, and straws that may contain the allergen and placed a high priority on the need to have dedicated dishes/cutlery to easily identify containers having the allergen-free food for the child (FARE 2016; Kim 2011). Awareness of the persistence of allergens in saliva has also been documented and is a cross-contact concern (Brough et al. 2013; Kim 2011). The potential for accidental transfer of minute amounts of allergen through saliva by others who had previously eaten allergen-containing food also reinforced the need to avoid sharing cups and utensils.

Restricting food consumption to the kitchen and dining areas helps minimize the spread of food allergens throughout the home and allows for effective and prompt cleaning of surfaces with general household cleaners (Perry et al. 2004; Watson et al. 2013) and preferably single-use disposable wipes or paper towels. Sponges are not typically recommended since they can also harbor bacteria and transfer microorganisms from one surface to another (Mirlei Rossi et al. 2013; Biranjia-Hurdoyal and Latouche 2016). If sponges are used, it is a good practice to frequently replace them and to have dedicated sponges for a specific purpose to minimize the transfer of allergen(s). Handwashing dishes thoroughly with warm water and dish soap or pre-rinsing dishes to remove visible residue prior to placement in a dishwasher is highly recommended (Brough et al. 2013; FARE 2016).

13.5   Communication, Education, and Emergency Care Training

Maintaining effective allergen controls within the home involves communication and education not only among family members and relatives, but also visitors, guests, babysitters, and others. Depending on the situation, it may be appropriate to tactfully share the importance of hand washing before and after meal preparation and dining. Communicating the family philosophy regarding the introduction of allergens into the home, whether it is in the form of a pot-luck meal, carryout food or edible gift, is important for the minimization of allergen transfer within the home. For example, one family experienced a situation where the babysitter came to the home eating a peanut butter sandwich, forgetting the fact that the children to be placed in the sitter's care had peanut allergies. The situation was resolved with a request for handwashing upon entry to the home and a review of the allergens of concern, as well as the emergency care plan.

The emergency care plan typically contains important information regarding signs and symptoms of an allergic reaction, instructions on how to administer epinephrine, phone numbers and procedures for calling emergency responders, and other emergency contact details in the event of a food allergic reaction. A variety of sample emergency care plans are available online for consumers to download and use (FARE 2017b). The completed emergency action plan with contact phone numbers is usually located in close proximity to the epinephrine and telephone. Some food allergic individuals also choose to wear a medical bracelet or tag to identify their health condition in case of an emergency.

It is critical that the allergenic individuals, family members, and primary caregivers be trained to recognize the signs and symptoms of an allergic reaction, know the location of the epinephrine and how to administer the medication via an auto-injector. Many families keep extra epinephrine in the kitchen or dining area where food is typically consumed and understand the need to use the epinephrine quickly and call 911 immediately in case of a reaction. Semi-annual or annual refresher training for the caregiver and the allergic individual is helpful, especially when the allergic child is old enough to self-carry and self-administer the epinephrine. The refresher training is also a good time to double-check the expiration date on the epinephrine and ensure that the epinephrine, antihistamines, asthma inhalers, and other medications have not expired.

Open communication between the allergist and patient/parent(s) is an important factor in determining the extent of allergen controls that may be needed. Prior allergic reaction history, other health concerns including asthma or eczema, and dietary or lifestyle preferences (vegetarian/vegan) also impact the general health and nutritional needs that must be met to enjoy a safe and healthy life. Involving dieticians/nutritionists and sharing experiences with other food allergic families through support groups and non-profit advocacy organizations provide additional forms of communication, education, and support.

13.5.1 Communicating Inadvertent Allergen Exposure

One of the new regulations issued in 2016 as part of the Food Safety Modernization Act (FSMA) requires a food facility to implement allergen controls. The final rule on Preventive Controls for Human Food covers many areas including the inadvertent introduction of allergens as a result of cross-contact. Along with establishing requirements for allergen controls, FSMA also strengthened FDA's authority to order a mandatory food recall (FDA 2016a). Published reports by Gendel et al. (Gendel and Zhu 2013; Gendel et al. 2014) have documented the continued rise in food allergen-related recalls to the Reportable Food Registry (FDA 2014b) and highlighted the high-risk food categories commonly involved in allergen-related recalls.

While both the Preventive Controls rule and FALCPA require reliable and accurate allergen information is provided on food labels to help protect the allergic population, consumers need to be aware that undeclared allergens are sometimes present in food products and can pose a risk to susceptible individuals. If a food product is suspected to have caused an allergic reaction due to an undeclared allergen(s), the following specific steps can be taken to help communicate information to the FDA after emergency care has been provided.

- If possible, keep the suspected food product in original packaging and record information or photograph the product label including product name, ingredient statement, lot number, and expiration date. Place the product in a secondary Ziploc bag and freeze the product if it is needed at a later date.
- Contact the FDA Consumer Complaint Coordinator for your region by phone or online; details for how consumers can report an adverse event or serious problem to FDA can be found on the FDA website (FDA 2016b), which describes four ways in which such voluntary adverse events can be communicated to the FDA:
  1. **Report Online**
  2. **Consumer Reporting Form FDA 3500B**. Follow the instructions on the form to either fax or mail it in for submission.
  3. **Call FDA at 1-800-FDA-1088** to report by telephone
  4. **Reporting Form FDA 3500** commonly used by health professionals.

In addition to contacting the FDA, the consumer may also want to report the adverse reaction to the manufacturer.

13.6 Summary and Conclusions

The decision to avoid or allow allergen introduction into a home is deeply personal and unique for each family. There are no "right or wrong" answers. The ability to adapt and modify the home allergen controls as a child grows or as the situation changes is key. Some families may initially have a lower degree of allergen controls in a home and may later modify it by increasing the safeguards if an anaphylactic reaction is experienced. Others may choose to have a high level of vigilance and attempt to avoid reactions to the greatest extent possible. Risk management for the food allergic individual, their families, and those who care for them is constant. Reading labels, effectively managing cross-contact risks, communicating, educating, and planning in advance for an emergency is a part of daily life for the food allergic individual and their caregivers.

Acknowledgements The author would like to acknowledge personal stories shared by numerous individuals during the Parents of Children with Allergies (POCA) of DuPage support group meetings, and various guest speakers. Conversations with other parents, food allergic adults, and their spouses are appreciated, with a special thanks to: Julie Hooven J.D., Roxana Dubash, Chris Powers, John Koontz, Ph.D., and Tim Duncan, Ph.D.

Disclosure Statements

No potential conflict of interest is reported by the author.

The author is a parent of two children with multiple food allergies and a member of FARE.

The opinions expressed in this chapter are the views of the author and do not necessarily reflect those of the U.S. Food and Drug Administration (U.S. FDA).

References

Biranjia-Hurdoyal, S., and M.C. Latouche. 2016. Factors affecting microbial load and profile of potential pathogens and food spoilage bacteria from household kitchen tables. *The Canadian Journal of Infectious Diseases & Medical Microbiology* 2016: 3574149.

Bock, S.A., A. Muñoz-Furlong, and H.A. Sampson. 2001. Fatalities due to anaphylactic reactions to foods. *Journal of Allergy and Clinical Immunology*. 107: 191–193.

———. 2007. Further fatalities caused by anaphylactic reactions to food, 2001-2006. *Journal of Allergy and Clinical Immunology*. 119: 1016–1018.

Brough, H.A., K. Makinson, M. Penagos, S.J. Maleki, H. Cheng, A. Douiri, A.C. Stephens, V. Turcanu, and G. Lack. 2013. Distribution of peanut protein in the home environment. *Journal of Allergy and Clinical Immunology*. 132: 623–629.

Crotty, M.P., and S.L. Taylor. 2010. Risks associated with foods having advisory milk labeling. *Journal of Allergy and Clinical Immunology*. 12: 935–937.

FARE (Food Allergy Research and Education). 2016. Creating a food allergy safety zone at home. Available at: https://www.foodallergy.org/file/home-food-safety.pdf. Accessed 27 December 2016.

———. 2017a. Dining out with food allergies:. Available at: https://www.safefare.org. Accessed 13 January 2017.

———. 2017b. FARE food allergy and anaphylaxis emergency care plan. www.foodallergy.org/faap. Accessed 24 January 2017.

FDA (Food and Drug Administration). 2004. Food Allergen Labeling and Consumer Protection Act of 2004 (FALCPA). Available at: http://www.fda.gov/Food/GuidanceRegulation/GuidanceDocumentsRegulatoryInformation/Allergens/ucm106187.htm. Accessed 12 January 2017.

———. 2013. FDA Guidance for industry: A food labeling guide - Ingredient lists. Available at: http://www.fda.gov/Food/GuidanceRegulation/GuidanceDocumentsRegulatoryInformation/LabelingNutrition/ucm064880.htm. Accessed 23 January 2017.

———. 2014a. Labeling of certain beers subject to the labeling jurisdiction of the Food and Drug Administration guidance for industry. http://www.fda.gov/FoodGuidances. Accessed 23 December 2016.

————. 2014b. The reportable food registry - A five year overview of targeting inspection resources and identifying patterns of adulteration: September 8, 2009–September 7, 2014. Available at: http://www.fda.gov/Food/ComplianceEnforcement/RFR/ucm200958.htm. Accessed 7 October 2016.

FDA (Food and Drug Administration). 2016a. FDA Food Safety Modernization Act (FSMA): Rules & guidance for industry. Available at: http://www.fda.gov/Food/GuidanceRegulation/FSMA/ucm253380.htm. Accessed 15 September 2016.

————. 2016b. How consumers can report an adverse event or serious problem to FDA. Available at: http://www.fda.gov/Safety/MedWatch/HowToReport/ucm053074.htm. Accessed 25 January 2017.

————. 2017. Consumer updates: Is it really "FDA approved?". Available at: https://www.fda.gov/ForConsumers/ConsumerUpdates/ucm047470.htm. Accessed 1 February 2017.

Ford, L.S., S.L. Taylor, R. Pacenza, L.M. Niemann, D.M. Lambrecht, and S.H. Sicherer. 2010. Food allergen advisory labeling and product contamination with egg, milk, and peanut. *Journal of Allergy and Clinical Immunology.* 126: 384–385.

Gendel, S., J. Zhu, N. Nolan, and K. Gombas. 2014. Learning from FDA food allergen recalls and reportable foods. *Food Safety Magazine.* April/May edition. 46–48, 50, 52, 80.

Gendel, S.M., and J. Zhu. 2013. Analysis of U.S. Food and Drug Administration food allergen recalls after implementation of the Food Allergen Labeling and Consumer Protection Act. *Journal of Food Protection.* 76: 1933–1938.

Glaspole, I.N., M.P. de Leon, J.M. Rolland, and R.E. O'Hehir. 2007. Anaphylaxis to lemon soap: Citrus seed and peanut allergen cross-reactivity. *Annals of Allergy, Asthma and Immunology.* 98: 286–289.

Gupta, R.S., E.E. Springston, M.R. Warrier, B. Smith, R. Kumar, J. Pongracic, and J.L. Holl. 2011. The prevalence, severity, and distribution of childhood food allergy in the United States. *Pediatrics* 128: e9–e17.

Hefle, S.L., T.J. Furlong, L. Niemann, H. Lemon-Mule, S. Sicherer, and S.L. Taylor. 2007. Consumer attitudes and risks associated with packaged foods having advisory labeling regarding the presence of peanuts. *Journal of Allergy and Clinical Immunology.* 120: 171–176.

Kim, J.S. 2011. Living with food allergy: Allergen avoidance. *Pediatric Clinics of North America.* 58: 459–470.

Laurière, M., C. Pecquet, I. Bouchez-Mahiout, J. Snégaroff, O. Bayrou, N. Raison-Peyron, and M. Vigan. 2006. Hydrolysed wheat proteins present in cosmetics can induce immediate hypersensitivities. *Contact Dermatitis.* 54: 283–289.

Lehrer, S.B., L. Kim, T. Rice, J. Saidu, J. Bell, and R. Martin. 2007. Transfer of shrimp allergens to other foods through cooking oil? *Journal of Allergy and Clinical Immunology* 119 (1, Suppl): S112.

Lehrer, S.B., S.W. Oberhoff, P. Klemawesch, L. Jenson, T. Rice, and S. Wunschmann. 2010. Unintended exposure to shrimp allergen: Studies of cooking oil used to deep fry breaded shrimp. *Journal of Allergy and Clinical Immunology* 125 (2, Suppl 1): AB226.

Mirlei Rossi, E., D. Scapin, and E.C. Tondo. 2013. Survival and transfer of microorganisms from kitchen sponges to surfaces of stainless steel and polyethylene. *Journal of Infection in Developing Countries.* 7: 229–234.

Muñoz-Furlong, A., and C.C. Weiss. 2009. Characteristics of food-allergic patients placing them at risk for a fatal anaphylactic episode. *Current Allergy and Asthma Reports.* 9: 57–63.

Perry, T.T., M.K. Conover-Walker, A. Pomés, M.D. Chapman, and R.A. Wood. 2004. Distribution of peanut allergen in the environment. *Journal of Allergy and Clinical Immunology.* 113: 973–976.

Pieretti, M.M., D. Chung, R. Pacenza, T. Slotkin, and S.H. Sicherer. 2009. Audit of manufactured products: Use of allergen advisory labels and identification of labeling ambiguities. *Journal of Allergy and Clinical Immunology.* 124: 337–341.

Pootongkam, S., and S. Nedorost. 2013. Oat and wheat as contact allergens in personal care products. *Dermatitis.* 24: 291–295.

Radke, T.J., L.G. Brown, E.R. Hoover, B.V. Faw, D. Reimann, M.R. Wong, D. Nicholas, J. Barkley, and D. Ripley. 2016. Food allergy knowledge and attitudes of restaurant managers and staff: An EHS-Net study. *Journal of Food Protection.* 79: 1588–1598.

Simons, E., C.C. Weiss, T.J. Furlong, and S.H. Sicherer. 2005. Impact of ingredient labeling practices on food allergic consumers. *Annals of Allergy, Asthma and Immunology.* 95: 426–428.

TTB (Alcohol and Tobacco Tax and Trade Bureau). 2012. Major food allergen labeling for wines, distilled spirits, and malt beverages. U.S. Department of Treasury. Available at: https://www.ttb.gov/labeling/major_food_allergin_labeling.shtml. Accessed 6 February 2017.

———. 2014. Consumer corner: Alcohol beverage labeling and advertising. U.S. Department of Treasury. Available at: https://www.ttb.gov/consumer/labeling_advertising.shtml. Accessed 6 February 2017.

USDA-FSIS (United States Department of Agriculture - Food Safety and Inspection Service). 2015. FSIS compliance guidelines: Allergens and ingredients of public health concern: Identification, prevention and control, and declaration through labeling. Available at: https://www.fsis.usda.gov/wps/wcm/connect/f9cbb0e9-6b4d-4132-ae27-53e0b52e840e/Allergens-Ingredients.pdf?MOD=AJPERES. Accessed 6 February 6 2017.

Watson, W.T., A. Woodrow, and A.W. Stadnyk. 2013. Persistence of peanut allergen on a table surface. *Allergy, Asthma and Clinical Immunology* 9 (1): 7.

Xu, Y.S., M. Kastner, L. Harada, A. Xu, J. Salter, and S. Waserman. 2014. Anaphylaxis-related deaths in Ontario: A retrospective review of cases from 1986 to 2011. *Allergy, Asthma, and Clinical Immunology.* 10: 38–38.

Chapter 14
Allergen Control for College and University Dining Service

Kathryn Whiteside, Lindsay Haas, and Marissa Mafteiu

14.1 Introduction

Food Allergy Research and Education (FARE) has recently created best practice guidelines for managing food allergies on a college campus. This is crucial because adolescents and young adults have the highest risk of food-induced anaphylaxis. The number of adolescents under age 18 with food allergies has increased by 50% between 1997 and 2011 (FARE 2014). These individuals may have help handling their allergies during primary education and while living at home, but are required to become autonomous when coming to college.

The eight major allergens that make up 90% of the food allergies in the U.S. are wheat, soy, milk, egg, peanut, tree nut, fish, and shellfish. The Americans with Disability Act (ADA) has recently declared that food allergies and celiac disease can qualify as disabilities because they put limitations on eating, a major life event (ADA 1990). Eating in the dining halls or retail cafes is an important part of a college student's social life, and every student should have access to safe food and be able to eat with their fellow students. The Association on Higher Education and Disability (AHEAD) encourages students to document their food allergies or celiac disease as a disability with the school to ensure that safe foods will be made available to them (AHEAD 2012). Managing food allergens is necessary to minimize the likelihood of adverse reactions when students dine at campus food-service establishments. Allergen control will be different for each school depending on the resources available to implement the protocols listed below.

K. Whiteside (✉) • L. Haas
Michigan Dining, University of Michigan, Ann Arbor, MI 48109, USA
e-mail: kswhites@umich.edu

M. Mafteiu
School of Public Health, University of Michigan, Ann Arbor, MI 48109, USA

© Springer International Publishing AG 2018
T.-J. Fu et al. (eds.), *Food Allergens*, Food Microbiology and Food Safety,
DOI 10.1007/978-3-319-66586-3_14

This chapter serves as a resource to other colleges and universities who strive to develop a policy to provide safe accommodations to their students living with food allergies or celiac disease. General best practices are described in regard to obtaining documentation of students' dietary needs, various serving solutions, necessary staff training required, and emergency response plans. Tracking allergens and gluten throughout procurement, storage, production, and service is also discussed, in addition to strategies to avoid cross-contact at all stages. Examples of resources used by the University of Michigan to provide students with allergen and ingredient information are illustrated as well.

14.2   Best Practices

Creating a system to ensure inclusivity and safety requires an interdisciplinary approach across campus. Published policies and procedures for how food allergies will be accommodated is one of the first steps that should be taken by the university to support diners. Dining, residential, disability and health services are involved in developing food allergy policies and distributing them to parents and students when a food allergy is reported. The policy should outline documentation that the student may need to provide, accommodations that can be made in dining and housing, staff training requirements, emergency response, allergy labeling provided, front-of-house and back-of-house procedures, and student responsibilities.

FARE held two college summits with key subject matter experts to help develop allergen standards for colleges and universities. There were representatives from over 50 schools in addition to delegates from AHEAD, the Department of Education, and other leading experts. Students and parents also attended and provided input into the development of these recommendations. These guidelines were published in 2014 (FARE 2014) and were piloted in 12 campuses across the country including University of Michigan.

The guidelines recommended that School policies regarding allergen management should provide the following broad requirements (FARE 2014):

- A campus-wide approach
- A transparent process to meet student needs without being burdensome
- A comprehensive food allergy policy
- Emergency response plans and training
- Confidentiality

The food allergy policy must be available across campus to staff and students, with online links to a centralized resource that provides the written policy, important contact information, and any required documents. The policy can be written within the university's disability policy or exist as a stand-alone policy that includes a process for students to declare their allergies and necessary accommodations. Staff training should be required so staff is aware of the policy and which components they are responsible for implementing. Training and implementation will vary based on the organizational structure of the campus and how the policy is written.

14.2.1   Documentation

Students can benefit from documenting their food allergies with the school so that the department overseeing allergen control is aware of their specific needs. The department can help facilitate communication between the student and personnel at the dining hall they are most likely to utilize during their stay. If the school has a Registered Dietitian, they can also meet with students to inform them of available resources and how to access safe, nutritious meals. Ultimately, it is up to the student to declare their food allergy, celiac disease, or other dietary needs to the university. There should be multiple opportunities for them to do so after being admitted, including when filling out housing applications or student health care forms, purchasing a meal plan, during orientation or open houses, on the school's website, etc. (FARE 2014). Other avenues to prompt students to disclose their dietary need may include campus tours for prospective students, social media, and printed materials sent to students. Giving students multiple chances to report their dietary requirements and providing knowledge of the resources on campus will increase the likelihood of documentation and subsequent accommodations for them while on campus.

AHEAD states that a university should evaluate a student's disability and accommodation needs on a case-by-case basis (AHEAD 2012). It is important to recognize that an equivalent disability may not require the same accommodations for other students. AHEAD advocates for documentation of food allergies or dietary restrictions to protect against discrimination and/or harassment (AHEAD 2012). Schools are legally obligated to accommodate students who register their food allergy as a disability, ensuring that they receive equal service and variable safe dining options. Food allergy documentation requires a statement from the student's physician within the last two years, as some food allergies can be outgrown. Celiac disease documentation should also come from a physician but does not need to be recent since it cannot be outgrown (FARE 2014).

The decision to require medical documentation of a student's food allergy or dietary restriction is at the university's discretion; while it is encouraged in case of emergency or to obtain disability services, the university can choose to serve all self-reported dietary needs equally regardless of the etiology. However, medical documentation may be required to exempt a student from a mandatory meal plan while living on campus. Medical documentation records should be stored with the department overseeing food allergy accommodations in a secure location and only accessed by designated personnel. Electronic backups of the records can ensure that they are not lost.

14.2.2 Accommodations

A central department (e.g., Dining services) should handle food allergy accommodations, the implementation of the policy, and serve as an advocate for students with food allergies on campus. However, representatives in other departments should aid in outreach to inform students of the food allergy resources on campus. Once a student's dietary needs are reported to the central department handling food allergies and accommodations, they should be encouraged to also register with the school's disability services department. This department can ensure that the student's professors are informed and necessary extensions are given in the case of food-related emergencies. Without proper documentation, the school might not have the authority to provide these services to students.

The food allergy policy will state who is responsible for implementing accommodations for students with food allergies or celiac disease and outline how this information will be provided to the student. Each student may have specific accommodations that fit their needs. Accommodations specifically related to student dining may include (FARE 2014):

- Identifying foods free of or containing allergens at every meal
- Developing procedures to avoid cross-contact
- Providing information on ingredients and food-prep methods to the student
- Giving the student access to an allergen-free preparation area
- Exempting a student from mandatory meal plans
- Housing assignments with a kitchen or student with the same allergy
- Academic flexibility if the student has an adverse reaction to a food

Serving solutions need to be considered carefully to avoid over-restricting some individuals by completely eliminating all of the eight major allergens or causing additional problems by providing limited solutions. For example, offering almond milk for individuals allergic to dairy will not help those individuals who are also allergic to tree nuts. Multiple solutions are best in order to accommodate a broad range of dietary needs. Specific accommodations that are currently employed in some universities include the use of dedicated allergen-free serving stations and prep areas, preordered meal service, a dedicated allergy-friendly or gluten-free pantry or a combination of these (FARE 2014). Descriptions and best practices for each are listed below.

14.2.2.1 Allergen-Free Serving Stations

These serving stations strictly serve dishes that are free of certain allergens that are decided by the school. These stations can be free of the eight major allergens or may be free of the most common allergens reported among students at a given college.

- These may work best when serving a large number of students.
- The school decides which allergens will be eliminated at these serving stations and implements effective back-of-house procedures to prevent inappropriate foods from being served here.

- The school decides whether the station will be fully accessible or limited to certain students. Limiting access to the station can lower the risk of cross-contact, but may also lead to more food waste. Students must be educated on their responsibility to prevent cross-contact of allergens at these serving stations.
- Ideally these stations will be conducive for a cook to fully prepare meals at the station. They will include ingredients stored in closed containers and separate cooking utensils. If a functional prep station is not possible, food may be prepared in a shared kitchen space, but procedures to prevent cross-contact must be implemented and carefully followed.
- Physical barriers between serving stations should be implemented to further avoid cross-contact, and to keep workers from easily crossing over to help other allergen-containing stations.
- Trained serving staff should be present at all times to prevent students from bringing outside food into the station. Proper signage and lists of banned ingredients should be kept at the station as well.

14.2.2.2   Preordered Meal Service

This service facilitates a relationship between the student and dining employees, allowing them to order meals catered to their dietary needs to be picked up at designated times from the dining hall.

- This may work well for smaller dining halls or when serving a small number of students with food allergies.
- Preparing allergen-free meals should be the responsibility of a designated chef or cook and should be done in a designated area free of the allergens being avoided.
- The cook should be given written documentation of each student's necessary accommodations.
- Ongoing communication with the student is helpful to ensure their satisfaction and reduce food waste.
- There should be a system for the student to easily contact the kitchen with enough time for them to prepare the meal before the student arrives for pickup. An online order form is ideal.

14.2.2.3   Dedicated Pantries

These are designated cabinets, refrigerators/coolers, or rooms that contain only food free of a given allergen or gluten. They can be accessible to all or require keyed access by permitted students with specific dietary needs.

- The school decides whether these will be open or closed access. Closed access facilities should be locked and only approved students will be granted entry. Pantries could require students to swipe their card for access or request a manager to unlock the room for them.

- Proper training about the use of utensils and equipment as well as prohibiting outside food in the pantry should be given to anyone with access. Training can be given online or in person. Signage should be used in the pantry as a reminder.
- Only specified foods are allowed and ingredient lists of all food in the pantry should be available to the students.
- All house-made food allowed in the pantry must be prepared in a designated allergy-friendly or gluten-free preparation area to avoid cross contact.
- Cleaning materials must be available for students to clean up after using the pantry.
- Pantries should include a refrigerator, microwave, toaster, counters, and cabinets. Additional equipment is optional.
- Stand-alone cabinets or coolers in the dining hall are an additional method of providing allergen or gluten-free foods and should follow the above guidelines as well.

14.2.3   Staff Training

Anyone involved with managing food allergies and celiac disease, regardless of their department should undergo proper training. It is important that dining managers, resident advisors, and everyone in between understand the severity of possible food allergic reactions and how to effectively prevent and respond to them.

Training should cover the basics of food allergies and celiac disease, symptoms of a reaction, and how to recognize and respond to a person experiencing anaphylaxis. It should also provide the university's policies and resources, and inform staff of continuing training requirements. Training will be specific to the employee's role at the university. For example, food preparers and servers will require more specialized training in regard to allergen and gluten control than resident advisors.

Many states have passed legislation regarding specific training regulations for managers of food service establishments. When developing training programs on campus, these regulations must be considered in order to ensure compliancy with the state as well as national requirements.

14.2.3.1   Resident Advisor Training

Resident Advisors (RAs) are often in closer communication with students than other university staff, and students may feel more comfortable approaching their RAs than other staff. This puts RAs in a great position to foster an open relationship with their residents and encourage them to practice safe behaviors. These include disclosing their food allergies or celiac disease to peers they live and dine with or those preparing their food, always carrying their epinephrine, and promptly seeking help for an allergic reaction.

It is also important for RAs to understand food allergies and be familiar with campus resources to guide students when necessary. When planning events, they should be cognizant of their residents' food allergies in order to provide an inclusive living environment.

Lastly, RAs should be aware of the risks of the presence of food allergens in alcoholic beverages. Alcoholic beverages are not federally mandated to list food allergens that may be present, and students consuming alcohol may forget to carry their epinephrine pens or carefully check foods before eating. Alcohol can increase the rate of food consumption and therefore the onset of allergic symptoms. Anaphylaxis and alcohol intoxication have many similar symptoms such as flushed skin, vomiting, confusion, and falling unconscious. RAs should always assume the worst when students with food allergies present these symptoms and should seek immediate medical attention.

14.2.3.2   Dining Staff Training

Dining staff should obtain detailed education about food allergens, gluten, and adverse reactions that can occur when proper procedures are not followed. Training should cover how to properly communicate allergen information and how to avoid cross-contact during food preparation and storage and cleaning procedures for both back-of-house (where food is prepared) and front-of-house (where food is served) (FARE 2014).

All dining staff should be directed on how to read food ingredient labels. The Food Allergen Labeling and Consumer Protection Act (FALCPA) requires all foods containing the eight major allergens to plainly identify them on the food label (U.S. Code 2004). There are two ways to state these allergens on a food label (see Fig. 14.1). They may be listed at the bottom of the ingredient list or within the list

| a) Ingredients: Enriched flour (wheat flour, malted barley, niacin, reduced iron, thiamin mononitrate, riboflavin, folic acid), sugar, partially hydrogenated soybean oil, and/or cottonseed oil, high fructose corn syrup, whey (milk), eggs, vanilla, natural and artificial flavoring) salt, leavening (sodium acid pyrophosphate, monocalcium phosphate), lecithin (soy), mono- and diglycerides (emulsifier). | b) Ingredients: Enriched flour (wheat flour, malted barley, niacin, reduced iron, thiamin mononitrate, riboflavin, folic acid), sugar, partially hydrogenated soybean oil, and/or cottonseed oil, high fructose corn syrup, whey, eggs, vanilla, natural and artificial flavoring) salt, leavening (sodium acid pyrophosphate, monocalcium phosphate), lecithin, mono-and diglycerides (emulsifier).

Contains: Wheat, Milk, Egg, and Soy |

Fig. 14.1 Two methods of stating the eight major allergens on food labels. (**a**) Include the allergen name within the ingredient list. (**b**) Include the allergen in a contains statement under the ingredient list (FDA 2004)

next to the ingredient that contains the allergen. It is important to illustrate these methods for identifying allergens on a food label so that they are not missed by those responsible for labeling food items.

Due to the ambiguous nature of precautionary allergen labeling on food products, schools need to create a policy on how to label within their facilities. There may be a designated person responsible for following up with food manufacturers to gain clarification for precautionary labeling on the products. The food allergy policy should be relayed to students so they are aware of how precautionary labeling is handled.

All staff should be trained to follow standardized recipes and avoid making substitutions unless the new product has an identical allergen profile to the old product. This will ensure the accuracy of the recipe's ingredient and allergen profile. Staff should also know the university's labeling policies (i.e., icons, allergens of concern, where to find information) and know how to direct students when they have questions in regard to food safety. The best policy in these cases is to designate administrative staff to handle all allergen or gluten-free concerns.

Dining staff is responsible for addressing any concerns they see during food preparation and to avoid cross-contact by following proper food safety and sanitation guidelines. Their job is to provide customer service to all students with special dietary needs and they should be willing to take extra steps to provide everyone with safe meals.

Front-of-house personnel need additional customer service training so that communication with students and other diners is consistent across campus. Regular training on policies and procedures for displaying allergen signage within the unit and menu design expectations should also be implemented. It is crucial for front-of-house staff to be knowledgeable about the ingredients used in all menu items and the procedures that back of house staff follow to prevent cross-contact in the kitchen so that accurate information can be communicated to diners.

14.2.3.3   Emergency Response

As students are most likely to experience an adverse reaction to a food where they live and eat, having emergency response policies in place and trained staff is essential. Worst-case scenarios of anaphylaxis require immediate epinephrine administration. Ideally, students will carry and administer epinephrine themselves, but the food allergy policy should state what to do if this is not the case. Questions to be asked when developing an emergency response plan are (FARE 2014):

- Are there state laws that may impact the campus epinephrine and emergency response policy?
- If/where epinephrine should be stored?
- Who should administer epinephrine and when?
- How far away is the nearest medical facility?

- Who are the campus emergency first responders, and do they carry epinephrine?
- What will be the standard procedures for seeking medical help?
- How will students' emergency contact information be obtained and recorded?

Since allergic reactions can escalate rapidly, it is necessary to create a plan that allows for immediate response in any situation. If the state and university regulations allow, designated staff should be trained in emergency response. Some universities will prohibit their staff from administering epinephrine to avoid liability; alternative solutions must be provided in the emergency plan to address this.

Emergency contact information for every student who reports a food allergy or celiac disease should be obtained at the time of reporting. This information can be stored by the department overseeing allergen management and needs to be easily accessible during times of emergency.

14.3   Allergen Controls: Labeling and Prevention of Cross-contact

Providing students with accurate ingredient and allergen information requires tracking allergens through every step of the food service process. This begins with procurement and continues through food preparation and service. Food allergy policies should strive to implement standardized recipes in all dining locations and incorporate a computerized database to store all item, recipe, and ingredient information for all food served on campus. A food service management system can provide soft ware that allows a university to manage their inventory, purchasing, service menus, item and recipe database, and nutrition information all in one system. Regardless of the system used, great attention to detail is necessary in order to properly identify exactly what is being served on campus.

The university can decide whether it will label foods as allergen-containing or allergen-free and whether allergens outside of the Big 8 will be labeled as well depending on the needs of the population (FARE 2014). Foods can also be labeled as gluten-free or gluten-containing, although it is crucial to be certain when labeling foods as gluten-free. As mentioned, the food allergy policy must include key strategies to provide ingredient and allergen information at each step in the food service process. Best practices for each step are listed below.

14.3.1   Procurement

This is the first step in allergen control and requires a contract between all vendors or suppliers and the university that describes policies in regard to allergen identification. Vendor responsibilities are (FARE 2014):

- The provision of accurate ingredient lists for all ordered products; all allergenic ingredients need to be identified
- Absolutely no substitutions for products unless prior approval is granted
- Informing the university when a product is no longer available or is known to have changed ingredient formulation

The university can include certain ingredients that they do not want to be served on campus at any time in the vendor contract as well.

Establishing a relationship with food distributors and manufacturers is helpful in case questions arise on how items were processed at the manufacturing plant. It may be helpful to distinguish whether something was processed on shared equipment with other allergen containing products or simply in the same facility but on a separate line. Additionally, cleaning and testing procedures of manufacturing equipment can minimize cross-contact, and thus should be described or documented.

14.3.2 Receiving and Storage

Items should be checked for damage and for risk of cross-contact during receiving. If cross-contact is suspected, the product should be refused. Additionally, the food allergy policy should dictate how unlabeled products are to be handled. Ideally, the supplier should be contacted to request ingredient information; if this information is not available, the items should not be used by the facility (FARE 2014).

Foods free from the eight major allergens, should be kept separate whenever possible during receiving and storing to minimize cross-contact with allergen containing foods. If space allows, designated shelves or rooms should store only products that are allergen free. Foods containing different allergens need to be segregated from each other to further reduce cross-contact risk and should be stored below allergen free foods. Products that are not easily contained (e.g., flour) should be kept in sealed storage bins.

Gluten-free foods should be segregated and stored in a similar fashion, ensuring that they are on separate shelves and stored above gluten-containing foods.

14.3.3 Food Preparation

Food production provides a risk of cross-contact unless proper policies and procedures are followed by trained kitchen staff. Allergy-friendly and gluten-free food preparation should ideally be done in separate areas with designated utensils and equipment. It is important to ensure that all equipment is washed, rinsed, and sanitized before and after prep. Designated equipment can be labeled or differentiated by color if desired. Staff should be alerted during allergen-free meal prep so that extra caution is taken to avoid cross-contact at this time.

The exact steps taken will vary by facility, but some additional practices to prevent cross-contact are (FARE 2014):

- Prepare allergy-friendly and gluten-free meals first to further reduce the risk of cross-contact.
- Prepare dishes free from specific allergens together (e.g., prepare all dairy-free items at once).
- Be mindful that steam or crumbs can contain allergenic particles and can easily transfer to allergen-free foods if they are not prepared separately from one another.
- Avoid carrying utensils or equipment that were used to prepare allergenic foods over a space used for allergen-free foods. Particles can drop and cross-contact can occur.
- If separate equipment is not possible, ensure that everything has been thoroughly cleaned to remove allergenic food residue before using to produce non-allergen containing foods.
- Cover all allergen-free dishes during baking to avoid contact with circulating allergen particles. HEAT DOES NOT DESTROY ALLERGENS.
- Do not add garnishes to allergy-friendly and gluten-free meals unless otherwise specified.
- Materials that cannot be washed, rinsed, and sanitized should not be reused for preparation of other meals (e.g., aluminum foil or cooking oil) unless they are used for the same allergen.

If a mistake is made during food production the product must be discarded and the recipe must be remade. Any amount of cross-contact could cause an allergic or adverse reaction.

All meals should be made according to standardized recipes so that all ingredients and potential allergens have been identified and substitutions must be avoided. However, the food preparer should pay attention to ingredient labels throughout production to ensure that hidden allergens are not present in the food.

14.3.4   Front-of-House Service

Providing an inclusive dining experience for students with food restrictions can be done by offering a variety of gluten and allergy-friendly dishes with accompanying signage. Signage with nutrition, allergen, and/or ingredient information of each dish can help a student decide whether or not it is safe for them to eat. Signage and labeling should be consistent across campus. The use of icons and bold print can make signage easier to read and increase students' confidence about the food. Ingredient lists must be accessible to students; if they are too lengthy to include on point-of-service signage, they must be available online, via mobile apps, or printed in the dining hall.

Signature			
Lemon Baked Cod	**M**	**gf**	❯
Arugula Pesto Sauce		**gf**	❯
Orzo & Bulgur Pilaf			❯
Zucchini and Summer Squash	**M**		**gf** ❯
24 Carrots			
Cheese Pizza	**M**		❯
Vegetable Jambalaya	**M**		**gf** ❯
Couscous with Parsley	**M**		❯
Olive Branch			
Beef Lasagna			❯
Fresh Steamed Broccoli Florets	**M**		**gf** ❯

●●●○○ Verizon 🔔   9:49 AM   88% 🔋▸

Beef Lasagna Nutritional Info

Sugar 3gm	7%
Protein 31gm	42%
Vitamin A 775iu	16%
Vitamin B1 0mg	2%
Vitamin B2 0mg	8%
Vitamin B6 0mg	8%
Vitamin B12 1mcg	16%
Vitamin C 3mg	5%
Vitamin E 0mg	1%
Calcium 500mg	50%
Folic Acid 10mcg	3%
Iron 2mg	11%
Magnesium 15mg	4%
Niacine 2mg	10%
Zinc 3mg	17%

Contains: eggs, milk, wheat/barley/rye

Fig. 14.2 Example of a mobile app used by Michigan Dining to provide nutrition and allergen information to students on the go (Reproduced from University of Michigan Dining 2016)

A mobile app was developed to show students at the University of Michigan web-menus and meal information on the go (See Fig. 14.2). Figures 14.3 and 14.4 illustrate point-of-service signage that provides either nutrition and allergen information or a disclaimer when this information is not known. Signage may be generated through the software that manages the item and recipe database. Management should always review signage for accuracy to catch potential errors in the database. Figures 14.5 and 14.6 are additional examples of web resources Michigan Dining uses to allow students to determine what foods will be compliant with their dietary needs at a given meal, and calculate the nutritional content if desired.

If ingredient lists are not available online or printed on location, students should be allowed to request ingredient lists and preparation methods for a recipe and designated personnel should be trained to answer these questions. Questions about food allergens or gluten should always be referred to these individuals. Answers should never be guessed at and only given when 100% certain. As mentioned, standardized recipes with documented ingredients and allergens should be available at every meal. Sometimes recipes are created to utilize leftover ingredients and may not be standardized. In these cases, a disclaimer sign should be put out by the dish to inform students that it has not been reviewed for ingredient accuracy (FARE 2014). It is better to have a label explaining allergen information is not available than to have an incorrect label posted. Students with food allergies or celiac disease should avoid dishes without allergen information.

Cross-contact can easily occur when students use the same utensil for multiple dishes, or spill ingredients in self-serve locations such as salad bars. Items containing the same allergen should be grouped together as much as possible (e.g., serving

Olive Branch

Chicken Shawarma [gf] [≋] [M]

3 oz Serving: Calories 192kcal / Fat 14gm / Sodium 159mg / Carbs 0gm / Sugars 0gm / Pro 15gm

Contains: milk

Fried Pita Chips [≋]

4 Chips: Calories 269kcal / Fat 15gm / Sodium 277mg / Carbs 28gm / Sugars 2gm / Pro 5gm

Contains: item is deep fried, wheat/barley/rye

Roasted Red Pepper Hummus [gf] [≋] [M]

1 oz Serving: Calories 40kcal / Fat 3gm / Sodium 126mg / Carbs 3gm / Sugars 1gm / Pro 1gm

Contains: sesame seed

Spinach & Red Onion Pizza [≋]

Slice: Calories 228kcal / Fat 6gm / Sodium 501mg / Carbs 33gm / Sugars 2gm / Pro 10gm

Contains: eggs, milk, sesame seed, soy, wheat/barley/rye

Pepperoni Pizza

Slice: Calories 243kcal / Fat 8gm / Sodium 566mg / Carbs 32gm / Sugars 2gm / Pro 11gm

Contains: eggs, milk, pork, sesame seed, soy, wheat/barley/rye

Deep fried items may contain additional allergens

Fig. 14.3 Example of a point-of-service menu sign that states allergens present in the food in addition to whether it is compliant with certain dietary preferences (Reproduced from University of Michigan Dining 2016)

different types of cheese on one end of the salad bar rather than mixed throughout the salad bar). Ultimately, students should be advised about the risk of cross-contact and be encouraged to request that a meal is made separately or taken from backup supplies that have not been exposed to possible cross-contact. Plated meals are another way to reduce cross-contact of allergens since they are served by dining staff instead of students.

When preparing separate meals for students with food restrictions, the server should repeat back the student's requirements after they order and before serving them. It is a best practice for the cook preparing the food to deliver it directly to the student; however, if an item is being made for later pickup or will be delivered by a

Fig. 14.4 Example of a disclaimer letting consumers know this recipe has not been reviewed for nutritional or allergen accuracy (Reproduced from University of Michigan Dining 2016)

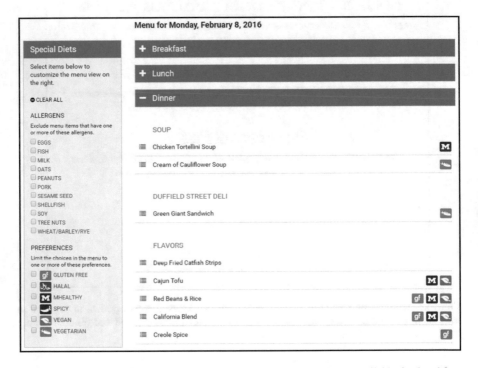

Fig. 14.5 Screenshot of a web menu at Michigan Dining. Menus are available for breakfast, lunch, and dinner at all dining halls and cafes on campus. Students can see the nutrition label by clicking on the item (Reproduced from University of Michigan Dining 2016)

server, a marker can be used to identify the allergy-friendly meal (colored frill picks identifying what it is free from, a label with the student's name, etc.). If the dish is not being prepared in front of the student and handed immediately to them, it should be covered to avoid cross-contact and individual, packaged condiments should be provided when possible (FARE 2014).

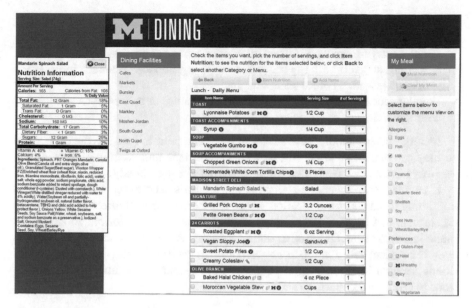

Fig. 14.6 Screenshot of MyNutrition resource at Michigan Dining. Students can filter the menu by allergy/dietary preference (on the right) and see a nutrition label with the full ingredient list (on the left) (Reproduced from University of Michigan Dining 2016)

While a dining hall can never claim to be completely allergen-free, a coordinated effort between all dining staff can provide accurate ingredient information and minimize the risk of cross-contact to provide students with a safe and inclusive dining experience. It is important for staff to be accommodating and know what resources are available for students with special requirements.

14.4   Student Responsibilities

By following the above guidelines, dining services can provide ample resources and information to students to help them make smart choices when dining on campus. It is also reasonable to expect students to fulfill some responsibilities, and that expectation should be clearly communicated to the student. Each college or university may have its own list of responsibilities it expects students to fulfill. Some examples might include:

- Disclosing a food allergy, celiac disease or other dietary requirement to dining services before arriving on campus the first time
- Filing emergency contact information with dining services
- A responsibility to check ingredient lists and read signage to make educated decisions about what to eat

- Communicating any dietary requirement before ordering a meal in the dining halls
- Always carrying epinephrine and other emergency medication with them
- Reporting any food allergy reactions to dining services (contact information should be provided before school begins)
- Voicing concerns regarding the number and quality of safe options available at their dining hall or other issues as they come up

Ideally students should have access to the menu prior to entering the dining hall so they can determine what will be available to them and coordinate with dining services if meals free from their allergen are not already available.

If an allergic or adverse reaction occurs, students need to report the incident to the dining manager at the location where it occurred, or to a direct contact in the department that handles food allergy control. This contact information should be made known to students prior to dining on campus. Incidents should be reported even if a student self-administers epinephrine to control the reaction. This will aid dining services in identifying the source of cross-contact and implementing solutions to avoid further harm to students.

14.5   Evaluation

The food allergy policy should include how it will be assessed for success. This may look different for every university, but will include a multidisciplinary team that will measure policy compliance and areas for improvement. Policy and process review can also be done by a third party. Performing self-audits of dining facilities can also help evaluate whether established policies and procedures for ingredient accuracy, signage and cross-contact prevention are being effectively implemented in the front-of-house and back-of-house.

Evaluation will ensure that essential components are included to properly manage allergies and celiac disease on campus. Criteria for success may include (FARE 2014):

- Continuing training for all staff involved in food allergen management
- Back-of-house and front-of-house policies and procedures
- Policies to establish accurate labeling and signage to help customers make informed decisions
- Continued meal plan purchasing of upperclassmen and sales of allergen-free and gluten-free items that show desirability of offerings
- Meeting your annual food service budget
- The absence of food-allergic reactions
- Inclusiveness of students with allergies or celiac disease
- Availability of a variety of allergen-free or gluten-free options
- Low student burden when reporting food allergies or celiac disease
- Easily accessible resources
- Students' perceived level of safety when dining on campus

Evaluating student satisfaction can be done by surveying students anonymously as they are more likely to provide honest feedback this way. Asking students about their experience may provide better data than simply monitoring incident reports since they may not document events that occur outside of the dining halls.

14.6   Challenges to Success

Implementing a successful food allergy policy can be made difficult by certain challenges presented throughout the management process. First, database management systems are costly and may not be suitable for small colleges and universities. Tracking ingredient and allergen information for all items and generating signage may not be feasible without these systems. Additionally, manufacturers may neglect to inform customers when product formulations change. Appointed staff should review products that are at high-risk for ingredient changes (i.e., processed foods, condiments) at set intervals and make necessary changes to the item database.

Creating an effective and comprehensive food allergy policy requires a great deal of labor up front. It takes time to train all appropriate personnel and ensure that kitchens and storage areas are able to accommodate allergen- and gluten-free meal preparation. Establishing an item and recipe database that will store information for every product served on campus necessitates trained staff with meticulous attention to detail. While it takes time to initially create the database, subsequent entries or modifications will be much simpler.

14.7   Conclusion

Coordination between many departments is necessary in order to implement an efficient food allergy policy on campus, but ultimately a central department should assume responsibility for accommodating students with dietary needs. Food allergy policies will be unique to each school and utilize resources or practices that work best for them. The best practice guidelines presented in this chapter were developed by experts in nutrition, food allergies, health and dining services directors, the U.S. Department of Education, and others. Many universities have implemented these recommendations and effectively provide access to safe meals for students with a variety of dietary needs.

References

ADA (Americans with Disabilities Act). 1990. Americans with Disabilities Act of 1990 as Amended of 1990 as Amended, 42 U.S.C. §12102. United States Department of Justice. Available at: http://www.ada.gov/pubs/adastatute08.htm. Accessed 1 January 2016.

AHEAD (Association on Higher Education and Disability). 2012. Supporting accommodation requests: Guidance on documentation practices. Available at: https://www.ahead.org/learn/resources/documentation-guidance. Accessed 1 January 2016.

US Code. 2004. Food Allergen Labeling and Consumer Protection Act of 2004 (Title II of Public Law 108-282, Title II). Available at: https://www.fda.gov/downloads/Food/GuidanceRegulation/UCM179394.pdf. Accessed 1 January 2016.

FARE (Food Allergy Research and Education). 2014. Pilot guidelines for managing food allergies in higher education. Available at: http://www.foodallergy.org/file/college-pilot-guidelines.pdf. Accessed 1 January 2016.

Index

Printed in the United States
By Bookmasters